The best design tips don't have to break the bank, as Sharon Hanby-Robie proved in her first successful home-decorating guide, *My Name Isn't Martha, But I Can Decorate My Home*. Now, she applies that same rule to home renovation. Learn from a pro:

- ❀ How to REALLY cost out a renovation without the add-ins from unscrupulous contractors
- ❀ How renovations will affect your property taxes and resale value of your house
- ❀ What questions to ask any builder, architect or designer BEFORE they start drawing up contracts
- ❀ What the most expensive part of any kitchen is . . . and it's NOT the appliances!
- ❀ When you can do it yourself, and when you should call in a pro
- ❀ Whether you can custom design your improvements and still stick to your budget
- ❀ What all those words that the plumber is throwing around mean

Everything you need to know is here, in one volume full of the information you must have to complete your home renovation, whether you're doing one room or the entire structure. With handy lists, cost-saving tips, shopping notes, inside design hints, and headache-sparing musts, it's the only book to read if you're ready for a new kitchen, bathroom, or second bedroom and want to plan ahead. . . .

MY NAME ISN'T MARTHA,™ BUT I CAN RENOVATE MY HOME

The Real Person's Guide to
Home Improvement

My Name Isn't Martha™, But I Can Renovate My Home

SHARON HANBY-ROBIE

POCKET BOOKS

New York   London   Toronto   Sydney   Tokyo   Singapore

Books by Sharon Hanby-Robie

My Name Isn't Martha,™ *But I Can Decorate My Home*
My Name Isn't Martha,™ *But I Can Renovate My Home*

Published by POCKET BOOKS

For orders other than by individual consumers, Pocket Books grants a discount on the purchase of **10 or more** copies of single titles for special markets or premium use. For further details, please write to the Vice President of Special Markets, Pocket Books, 1230 Avenue of the Americas, 9th Floor, New York, NY 10020-1586.

For information on how individual consumers can place orders, please write to Mail Order Department, Simon & Schuster Inc., 200 Old Tappan Road, Old Tappan, NJ 07675.

An *Original* Publication of POCKET BOOKS

POCKET BOOKS, a division of Simon & Schuster Inc.
1230 Avenue of the Americas, New York, NY 10020

ISBN: 0-671-01543-5

First Pocket Books trade paperback printing August 1999

10 9 8 7 6 5 4 3 2 1

POCKET and colophon are registered trademarks of Simon & Schuster Inc.

Cover design by Jeanne M. Lee
Front cover author photo by Steve Skoll; other photos by PhotoDisc
Text design by Stanley S. Drate/Folio Graphics Co. Inc.

Printed in the U.S.A.

ACKNOWLEDGMENTS

My many thanks to all those who have allowed me the pleasure of working with them to improve their homes. I am also thankful for all the fine craftsmen, carpenters, cabinetmakers, and contractors who have taught me and helped make the expression of dreams a reality.

I thank my parents for their continued encouragement and words of wisdom.

Thanks to all my friends for their loyal support and prayers.

Special thanks to my husband, Dave, whose continued faith and encouragement is a constant source of love and support.

Praise and glory to God, the inspiration of all creativity.

CONTENTS

EXPLANATION OF ICONS

Throughout the book, there are many useful asides meant to guide, give caution, suggest ideas, and provide advice and pertinent information. These sidebars are divided into the following four categories:

 TRICKS OF THE TRADE—Advice and inside information about the world of home renovating, revealing industry secrets on how to cut costs, discover hard-to-find bargains, and much more.

 FISTFUL OF DOLLARS—Facts and figures you need to know in order to stay on budget and save mounds of money while renovating your home.

 IDEAS AND TRENDS—Like your own personal interior designer, these sidebars will suggest helpful hints on the latest in decor, fashion, and selecting the perfect styles to suit your own tastes.

 BE FOREWARNED—Essential cautions on how to side-step pitfalls, avoid getting ripped off, and be prepared for every renovating contingency that may arise.

INTRODUCTION

My Name Isn't Martha,™ but I have had a lot of experience helping people improve both the comfort and value of their homes. With a little education and some words of wisdom, even Martha would be impressed with what you can do to better your home. As an interior designer, I have the advantage of having learned from years of experience what it takes to do things right. And fortunately for you doing it right doesn't necessarily mean doing it the most expensive way. When my editor asked me to do a book on home improvement, I began searching through my files for information. I found an interview from a newspaper article I did in 1991. It was interesting that most of what I said then is still true today: many people are opting to renovate rather than move. What makes this interesting is that for the most part, you can build a new home for about the same price as redoing your old home. So why choose to renovate? This is the most important question you can ask yourself.

Gayle Butler, editor of *Remodeling Ideas,* says: "Remodeling a home is a way to protect our roots, while reshaping our homes to new needs." I agree with her completely—I believe that we need some constants in a world that has us juggling to keep up with the hectic pace. We all crave a closer relationship with family and neighbors. We often feel isolated in our own little world. We are tired of starting over, and we want a place we can truly call home—not just a roof over our heads. When we find a place we find comfortable, we want to stay.

My desire is to give you the information you need to accomplish this goal, whether you are redoing an old charmer or adding charm to a new home. I will take you through the reality check of value versus comfort. I will help you decide what is best and most appropriate for your specific situation. Soon you will also discover how wonderful life can be in a home that is designed just for you.

I am sure you have already heard some disaster stories associated with home renovation. But don't believe everything you hear. With the

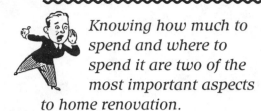

Knowing how much to spend and where to spend it are two of the most important aspects to home renovation.

right information and some key tips, improving your home ultimately can be a rewarding experience, rather than a punishing one.

This book is divided into three basic sections: Evaluating and Planning, Interior Renovations and Products, and Exterior Renovations. To make it even easier, each section is divided room by room. Throughout, you will find budget options and other financial considerations. The question my clients ask me, as a designer, most often is, "What can we do to improve our home?" Many people can tell you *how* to do the renovation, but few tell you *what* to renovate.

So grab your wish list, and let's get started!

PART ONE

EVALUATING AND PLANNING

1

TO MOVE OR IMPROVE—
THAT IS THE QUESTION!

As both an interior designer and a real estate agent, I have the advantage of being able to consider home improvement from two different perspectives. As a designer, I want to be a problem solver as well as to aesthetically improve the appearance of a home. But as a real estate agent, I want to do what makes the most sense from an investment standpoint. As a result, I often find myself trying to resolve conflicts resulting from these two perspectives. So I have compiled a list of questions to ask yourself in order to determine what is the best solution for your specific situation. The answers will obviously be determined by you, your family, your lifestyle, and your future dreams and plans. I have offered commentary to the questions in order to further help you in making your decisions to renovate.

❀ *How long do you anticipate living in your home?*

Certainly, plans can change. Yet it is important that you realistically answer this question as best you can. If you decide to proceed with your renovation plans but you also intend to move in a few years, then I may advise you to do things differently than if you were planning on staying in your home for many years. Unless you are planning on living in this home for the rest of your life, it is necessary to give some consideration to resale. At some point, most of us will sell our homes. It should come as no surprise that the homes that take the longest to sell are those that are overpriced, overimproved, and far too personalized for the general public.

A prime example of this is the home across the street from me. Even I could not sell it, and I tried very hard. The problem is threefold.

The foyer and dining room floors are black and white granite. Inlaid in the center of the foyer floor is an enormous black granite star. The living room and family room fireplace facades, which extend all the way to the ceiling line, are *pink* and white marble. The two columns in the family room are also the same pink and white marble. The kitchen countertop is yet another black-, gray-, and rust-colored granite. All of this marble was not only very expensive, but it is also completely out of character for the rest of the neighborhood. And maybe even the rest of the area. As a result, it has put the cost or value of the home far above the others when you consider in the balance the available amenities. Second, the marble touches are highly personalized and do not fit the tastes of many home buyers who have been looking in the neighborhood. Third, our neighborhood is new and still under construction. Which means you can still build the home of your dreams from scratch rather than purchase an existing home on our street. The family that built this marble interior home had planned to live there a very long time. Unfortunately circumstances changed suddenly, and they found themselves moving back to Italy. Their home has now been for sale (and vacant) for two years.

✾ *How does your home compare in price to the rest of your neighborhood?*
This is one of the most important questions you can ask yourself. If at all possible, you do not want to have the most expensive home on the block. The most expensive home is reduced in value by the average price of the others on the street. In the real estate market, only the last twelve-month sales numbers count. Most homes are valued by the CMA process— comparative market analysis. This is an average of the price that homes of similar style, size, and location sold for in the past year. This information is public record at your local courthouse. It is also available through any real estate office.

Ideally, you would like to be the least expensive house on the block. The reason is simple: the higher value of the other houses in your neighborhood automatically increases your house's own value. Unfortunately, the opposite is also true.

✾ *How does your home compare to the others in amenities?*
In other words, how many bedrooms do most homes on your block have? Do they all have basements, attics, garages, large lots,

and swimming pools? Or are you the only one considering adding an extra special amenity? Will you end up having the only home with a family room, but no garage or basement? If so, you may have trouble selling. My sister is currently finding it difficult to sell her home. She has one of the few homes in her neighborhood without a basement. It didn't matter to her when she purchased it. However, now that there are eleven homes for sale in her neighborhood, it seems to be a big deal to everyone else!

❀ *Why have you decided to improve your existing home rather than move to a new home?*

There are many valid reasons for staying and improving your home. As I said in the introduction, protecting your roots is a very good reason. Speaking of roots, yards are very high on the list of reasons to stay put. Our yards are the windows to the heart of our home. We take great pride in grooming and landscaping them. It's as if our yard is God's little acre to us, and we become very attached to them. Oftentimes, our reason for renovating may be to satisfy a desire for some-thing *different*. One of my clients nearly doubled the size of their home, knowing full well that it would price them way out of the neighbor-hood average. But they loved their neighbors, and their children were extremely happy. So this renovation made sense for them. Even after a fire destroyed their home, they rebuilt. When making your own deci-sion, be sure to weigh the pros and cons, based on your own desires and needs. I usually suggest to a client that they at least do a little house window-shopping before actually signing a contract to proceed with a major renovation. It just makes good sense to see what else is available. Most people also find it educational in terms of making good choices and decisions for their own home project.

❀ *How much money will it cost and does it make sense?*

Cost is certainly a valid consideration to renovating versus mov-ing. Most people will spend 20 percent more than the value of their current home for a new house. Then you have to add another 20 per-cent for commissions, attorneys' fees, and moving costs. That doesn't take into consideration the money you'll also spend to prepare your old home for "sale," or the cost of getting the new home to meet your standards and requirements.

Remodeling is not necessarily less expensive, but you may be able to get more for your money than you would by moving. One reason is that the money you would spend on moving can just as easily be spent on improving your existing home. One consideration in remodeling that people forget about is how it will affect their property taxes. One of the reasons a building permit is required for home improvements is to keep track of them and assess them for taxes. As a result, you will have an appropriate rate of additional tax added to your yearly tally. Be sure to include this when calculating your budget for the project. And don't forget the aggravation of living through a renovation! One of my clients became crazed when she realized that we would have to turn the water off completely while renovating her kitchen. She was also surprised that even though we had used heavy plastic to shield the other rooms from the work, dust still accumulated throughout the house. On the other hand, another client took a total house renovation in stride—and she had two teenage daughters living at home! They simply moved to the basement for six weeks. They prepared a minikitchen with living and sleeping quarters and decided to enjoy the ride. The difference was in their personalities. Knowing how you will deal with renovation aggravation is important to your decision.

The average cost of adding a family room is $31,000, a major kitchen remodel can cost at least $21,000.

A poll conducted by the National Association of Realtors in 1996 listed the following top ten reasons for moving:

1. 83 percent wanted a house they liked better within their price range
2. 79 percent wanted to live in a better neighborhood
3. 46 percent wanted more space
4. 41 percent wanted to live closer to their work
5. 38 percent wanted to send their children to better schools than were available in their neighborhood
6. 23 percent wanted to be closer to friends and family
7. 18 percent wanted to be closer to parks and recreational areas
8. 17 percent wanted shopping areas close by
9. 12 percent wanted to live in a planned community
10. 5 percent wanted better access to public transportation

Unfortunately, to really make an informed decision about whether to move or renovate, it will practically require going through the entire process of planning the renovation to properly make a choice. In a nutshell, you will need to talk to real estate agents, study property values, and talk with remodeling contractors. You will also need to make precise lists of all specifications and finish materials, including appliances and fixture choices. Finally, you will need to get a contractor to give you a rough estimate for your plan. Below is a table that will provide some insight to cost averages for the most common home-improvement projects. It also includes the added investment value used for selling a home and the percent of gain recovered. This can be helpful to making a decision to move or stay.

You will not get 100 percent of your renovation investment cost back when you sell your home.

BUDGETS—THE COST OF REMODELING

Project	Cost	Resale Value (National Average)	Cost Recouped
Family room 16 × 25 ft.	$32,558	$27,904	86%
Two-story addition 16 × 24 ft. w/bathroom 5 × 8 ft.	56,189	48,943	87%
Kitchen remodel	22,509	20,340	90%
Bathroom remodel	8,563	6,582	77%
Master suite (w/bath) 24 × 16 ft.	37,388	32,527	87%
Deck addition (16 × 20)	5,927	4,356	73%
Replace siding	5,099	3,593	71%

Source: "Cost vs. Value Report," *Remodeling*, November 1997, (Figures based on results of 1997–1998 report)

So Where Do You Begin?

Let's assume you are fairly certain you want to stay where you are and renovate. As I suggested earlier, do some house window-shopping first. Take several weekends to attend open houses in new developments. This will give you an opportunity to see what the newest trends in your community include. Home shows are another good resource for finding out about new products and contractors. It gives you an opportunity to interview several contractors face-to-face, to see how your personality and style fit with theirs.

By now you should be beginning to develop an idea of what you want. It's time to begin drafting your rough list of "wishes." Be as precise as you can. I usually suggest two lists. One that details all of your goals, and another to address specifics like size, location, etc.

An example of the "goal" list would be: a place for a computer, more storage, a music space, a home office, or more room for the children to play. The second list could include more specific things such as: a larger kitchen with an eat-in area, new cabinets and appliances, a larger master bath, or adding a garage or a screened-in porch. Sometimes we just know we need more space, but we are not sure exactly what kind of space would be most effective, or where to put it.

Whom Should You Consult?

The answer to this question will depend on whether you are a DIY (Do-It-Yourself) person or an HAP (Hire-A-Professional) person. If you are a qualified DIY person, then consider the size of the project and the amount of time involved to complete it. You may want to break the project down into bite-size pieces to determine exactly what you can or are willing to do yourself. Even the most qualified DIY person will need to work with good resource people. My mother has been in the kitchen and bath remodeling business for years. Her knowledge and instincts combined with her experience make her a very good choice as a resource. She works for a well-established company that supplies bathroom and kitchen fixtures (sinks, tubs, etc.) to both the retail and wholesale trade. In addition, they have cabinetry available for most areas of the home. The knowledge and information available from a company of this nature is invaluable for the DIY person. With hundreds of catalogs at their fingertips, they can supply information on such things as brand differences, sizes and specifications, installation instruction, and maintenance issues. If you are a DIY person, then a reputable firm of this nature can be a tremendous help.

If the scope of the renovation requires a project manager, and you are thinking about acting as the manager yourself, then considerable thought must be given to the realistic possibility of you being able to handle the project without it interfering with your normal workload. A general contractor can help work your own ideas into a practical plan for construction. Many contractors are quite capable at designing creative and economical ways of doing most anything. In most cases, they are willing to work with you in such a fashion as to allow for your own participation in some of the work. The contract would need to be written with specific language that spells out exactly who will be doing what. I'll go into further detail on choosing and working with contractors later in this chapter.

If you are not the DIY type, then the next step in the process should be to get a professional involved to help you. A professional can be one of several people: an architect, a designer, or a contractor. I prefer using a designer or an architect as my first reference point. These are what I consider the "idea" people. Because they will not profit from the actual construction, they can be more objective and attentive to your structural and financial needs. However, I know several contractors who either employ designers full-time or bring them in as consultants when needed. One of my favorite builders often hires me on a single-project basis. I have found working with his staff to be a good experience. Of course, I do have an advantage—I am a designer. Which may make it easier for me to understand certain concepts and details than someone without the experience or visualization skills of a designer. Nonetheless, the builder's home designer has been very helpful and creative with design plans and solving problems. The job of a designer or architect is to help you determine and visualize how to meet your goals in the most effective way. They have the ability to give you options that will ultimately allow you to work with a couple of different budgets. Because they have enough knowledge about products and specifications, they can guide you to low-, mid-, and high-end ways of doing things.

How do you locate these people? I always suggest getting recommendations from friends, neighbors, and family. The Better Business Bureau can give you a list of local members, as can the Building Industry Association (BIA). Ultimately you should still interview and check references of anyone you are considering. One of the key points to consider is whether or not you are compatible personalities. No matter how accomplished a designer is, if you do not like her, she cannot please you! I recommend interviewing at least three people. At least one should be an architect. When first calling to set up an interview, ask how large or small a project do they usually consider. Also, be sure they specialize in residential construction and home improvement. You do not want someone whose area of expertise is in commercial construction who just happens to need some extra work!

Here is a list of some basic questions to ask:

1. How do you determine your fee schedule? Is it flexible?
2. Will you "manage" the project, and if so, how is this priced?
3. How long have you been involved in renovation? Do you have a portfolio?
4. Ask for a list of references and diligently follow up on several of them (see below).
5. Ask if they have ever had a client bring a suit against them. If so, why? How was it resolved?
6. Will you be involved in recommending and selecting materials, appliances, fixtures, and finishing products?
7. Will you provide a detailed schedule for completion, and drawings for the contractors to bid on?

Remember, *you* are in charge. You will ultimately determine how involved you want this person to be. You may want them to provide only ideas while you do all the legwork and manage the project. On the other hand, you may want to have them help with every aspect of the job. Make your desires clear right up front. Communication is the key to any good relationship. I have worked with clients on many different levels. Some only needed me to confirm their own ideas before proceeding with a contractor. Others have asked me to design, specify materials, recommend contractors, acquire bids, and manage the project completely. The choice is yours.

The next step should be to check references. Here is a guide to some questions to ask:

1. Was she easy to get along with? Was she comfortable working with both you and the rest of your family?
2. How big a job was it?
3. How was your working relationship? Do you feel she cooperated with you and your desires on the design plan?
4. Did she give ample attention to details such as needs, costs, and design decisions?
5. Did she use a contract? How did she determine the fee schedule?
6. How was her relationship with the contractor? Did the contrac-

tor have a lot of complaints with regard to the design and ease of implementation?

7. Have there been any unforeseen problems with the overall plan or design now that it's finished? Has she been willing to address them?
8. How accurate was she in terms of timing?

Once you have chosen a designer or architect here are some rules for you to follow:

Be honest about your needs and budgets. John Rusk, author of *On Time and On Budget,* recommends running numbers through your head to determine your own comfort level. Start with $10,000 and continue in increments of ten such as: $20,000, $30,000, $40,000, etc., until you reach a level where you are *uncomfortable*. This will give you a ballpark range for discussion. Knowing this number is critical to any discussion on home improvement. Using this number as a temporary budget when beginning your design plan can be the key to accomplishing your goals in a realistic manner. After all, it is possible to renovate a kitchen for $15,000 or $50,000. Knowing what you are willing to spend will make a huge difference in how a designer approaches the plan. Remember, this number needs to be acceptable in terms of how it affects the overall value of your home.

Don't pretend to be something you are not. Designers and architects are just like you. They have families, watch TV, their kids track mud through the house, and they leave their ironing board up in the living room. In other words, they live in the real world too. But if you suspect at all that the ideas and/or plans they are suggesting are not consistent with your lifestyle—stop them! You do not want a room that looks like a magazine photo, you want something that will make your life easier and better! There are basically two types of designers: ABC designers (my terminology) and creative designers. The ABC designer basically prefers to work within only specific styles and/or price ranges. For example, I know a designer who specializes in eighteenth-century styling. Unless you and your home *fit* his plan, he can't help you. In contrast a creative designer's goal is to get to know you and your family and all of your needs and design accordingly. How do you know the difference? If they immediately start coming up with ideas before they have even interviewed you or taken the time to think about the project—be careful—they are probably an ABC designer!

The Contractor

There are probably more horror stories about contractors than about any other professional. We have all heard how "contractor impostors" prey on the elderly and take them for thousands of dollars for repairs that not only were *not* necessary but were then billed at ten times the real cost. The truth of the matter is that this is the easiest problem to avoid. Never, ever, sign a contract with someone who contacts you first. If you didn't call them and ask for their opinion or an estimate, then BEWARE!

The choice you make and why you make it often determines the kind of experience you will have. As always, you get what you pay for. As with the designer/architect, getting recommendations is always the best bet. If your home is less than ten years old and you are happy with the overall construction and quality, then I would consider the original builder, if he is still available. The reason is simple: he knows exactly what was done and where things are located. Otherwise, once construction begins, you may discover a few surprises, for example, finding a steel support column where you had hoped to put a doorway can throw a real monkey wrench into your plans.

Another way to find builders is through new home development projects in your community. Choose a home that you find attractive and simply interview that builder. Many times people forget about home builders and look only to renovating contractors. Builders are a wonderful resource for renovation. Because of the volume of work they do, they can often be very competitive in pricing, particularly on materials such as windows, doors, gas fireplaces, etc.

That old wives' tale, "Nobody builds a quality home any more," is just that—an old wives' tale. Ultimately, you will need to determine how much you are willing to pay for style and quality. Some contractors/builders are able and willing to work within different levels of cost and quality. Others may not be willing to compromise their reputation and therefore won't be comfortable working below a certain level of established quality. As a result, the process of interviewing a prospective contractor should involve not only *talking* with him and *examining his work in person,* but also *checking out his references.* As mentioned earlier, it is essential to like the people you are working with. No matter how competent someone is, if you have a personality conflict with

If substitutions for a certain product or service are suggested, be sure that they are equal in quality to the original. Some builders may have an ongoing relationship with a competitive manufacturer with the same quality materials. It would be a cost advantage to use this source rather than opening a new account.

them then they will not be able to make you happy. Communication is the most important tool you have to create a successful project. Do not expect to get anything you have not discussed. Detail, detail, and more detail is the key to everyone being satisfied.

Once you have an established plan, you may have an opportunity to have contractors bid for the work on your home. I recommend getting at least three bids for every project. The most important aspect of getting comparable bids is being sure your specifications are complete before the bidding process begins. If you have worked with an architect or designer, then let them do the specification list for the bid.

CONSIDERATIONS AND QUESTIONS FOR THE CONTRACTOR

1. Do you use standard AIA (American Institute of Architects) forms? Form A107, which is the Abbreviated Owner-Contractor Agreement and is used by small construction contractors, is endorsed by the Council of General Contractors as well as the AIA. This is an advantage, particularly if you are using both an architect and a contractor, because they will both be comfortable and familiar with the terminology and requirements of this contract. It can be obtained at website: www. buildersbook.com, for a fee of four dollars.

2. Do you provide detail specifications for quality control? In other words, get in writing details including brand specifications and model numbers of all products to be used.

3. Does your contract provide for conflict resolution? If so, how? *Mediation* is my first choice if you cannot work it out yourselves. (See Words of Wisdom, later in this chapter, for further information.) An example of such a situation is when I had my screened-in porch built last fall. The building inspector said the railing did not meet the building code standards.

For safety reasons, it should have been eight inches higher. Since the builder had drawn all the plans and obtained the necessary permits, I had not specified the measurement of the railing but simply chose from the samples the builder provided. I felt the error was not my responsibility but the builder's. It was an expensive mistake, and I was fortunate because my contractor readily admitted that it was his problem, not mine. The original railing had to be ripped out and replaced before I could use the porch. If the builder had not admitted to being responsible for this error, it could have been very difficult to resolve. Planning ahead and obligating both parties to a specific method of resolution can save not only money, but a lot of stress and arguing.

4. Will you contract a finish date? Usually you will not get a contractor to agree to a penalty for not meeting such a date. However, if you are willing to give a bonus for early completion, then most contractors will agree to a contracted completion date.

5. Discuss liability issues and ask for copies of their insurance coverage. Have it checked out by your attorney and abide by his suggestions.

6. Is the contractor a member of the BBB (Better Business Bureau) or BIA (Building Industry Association)? If not, why?

7. Obtain a payment schedule and have it reviewed by your attorney. This schedule should provide for reasonable holding of funds until completion of the project. That way you and the contractor will be obligated to this schedule. This eliminates inappropriate demands for money. Sometimes a contractor can get behind in his accounts payable. This can make it difficult to convince subcontractors who have not been paid to show up for work. This is not your problem but the contractor's.

8. Do you use subcontractors? If so, get their names and references. Oftentimes, the price and quality of your job is determined by the price and quality of the subcontractors employed.

9. Discuss job maintenance. Maintaining a clean, organized site is imperative to a well-executed job. Be specific about your desires and needs. If you have children in the home, you must

be assured of their safety after the crew has left the site for the day.

10. Who will be responsible for obtaining necessary permits, inspections, filings, and engineering? Are the fees associated included in the contract? Since this can be a confusing aspect of the project for a novice, I prefer that the contractor/builder be responsible for acquiring necessary permits.
11. Discuss how timing and pricing will be affected in the event of any necessary changes after construction has already begun.

Basically, all of the above situations should also be included when you question references. How situations were resolved with previous customers will give great insight into what you can expect. In addition, the most important question to ask a reference is: Were your expectations met? The most dissatisfied customers either didn't plan well or didn't understand the plans. I cannot stress enough how important it is that you are crystal clear on every item and detail. It is your responsibility to know exactly what has been specified. And the only way to know this is to actually see it.

Here are some other areas that are the responsibility of the owner:

1. Daily, or at least weekly, inspect the job site. Be sure doors, walls, windows, switches, and lights are where you expected them to be.
2. Report any concerns and/or changes.
3. Read any change orders, including the fine print. Be sure a price has been included and that you exactly understand what it entails.
4. Ask questions—even if you know the answer. It will keep the contractor on his toes if he knows you are watching.
5. Don't jump to conclusions. Some things will just not look right until they are completely finished.

Should I Borrow or Pay Cash?

Once you have established the parameters and basic budgets, you need to decide how to pay for this project. In many ways this will be determined not only by the amount of available cash you may have, but also by your personal viewpoint on borrowing, and how long you plan on staying in your home. Recognize first that a home equity loan's interest payments are tax deductible. If additional tax breaks are an advantage for your situation, then this may make the most sense. There are three different types of home-equity products available: a *conventional* loan—also known as a *second mortgage;* a *home-equity line of credit*—which is a revolving line of credit, which you can borrow from at your own discretion, and *refinancing*—which replaces your existing mortgage. Regardless of the type of loan you choose, the amount you can borrow will be based on how much equity you have in your house. In most cases you can borrow up to 80 percent of the equity. There are some lending institutions that will allow you to borrow as much as 125 percent of the equity of your home. The problem with this is that it makes it virtually impossible to pay off. Besides, the higher the amount you borrow, the higher your interest rate will be.

One word of caution, do not let any lender talk you into their "bimonthly" payment plan. This is something you can do yourself without charge! It is a great way for lenders to make money, since they usually charge $300 for this "service." When I questioned a mortgage lender on this, his answer was that most people are not disciplined enough to do this on their own. Therefore, if they are paying for such a privilege, they will be consistent with paying. In other words, do it yourself. You decide when and how much you desire to pay additionally against your principal loan.

If you have cash, you may be able to earn more investing the cash than it will cost you to borrow on interest, especially with the low rates that are available today. How? With interest rates as low as 7 percent for a home-equity loan, you can invest in a mutual fund that may earn twice as much. Compounded over the term of your loan, you can really accumulate a nice nest egg. In addition, you will save on your taxes by deducting the 7 percent interest you're paying on the loan. This is a great deal!

If you have not refinanced your home for some time, then this can be a real viable option because of the increased equity that may have accrued. Equity is the difference between the market value of your home and your existing mortgage debt. Of course, this too will be dependent on current interest rates compared to your existing mortgage loan rate. Often if the value of your home has risen, it may allow you to spend more than you thought while still keeping your mortgage payments at their current level. Obviously this can mean extending the mortgage term. But the benefit may be worth it. In my particular situation—being in that "midlife" stage—my accountant highly recommended that I invest any available cash into long-term retirement vehicles such as US Treasury strips, and stocks. This meant that I ended up with a higher mortgage than I had planned, but the deductible interest combined with the growth of the retirement investments more than make up for my desire for a smaller mortgage loan. There is also a new mortgage available called the HomeStyle mortgage from Fannie Mae. It is based on the home's appraised value *after* renovations—not on its value at time of application. This allows you to borrow enough money to finance major repair work. For more information, call Fannie Mae, 800-732-6643.

If the thought of extending the term of your mortgage really bothers you, consider the fact that you can add extra payments toward your principal as often as you like. This will in effect reduce the term and overall interest of the loan without obliging you to a shorter term.

Words of Wisdom

❀ Before construction begins, discuss and agree in writing to a plan for conflict resolution. Again, I suggest that *mediation* should be the first course of action. It does not require attorneys and is a fair solution for both parties. In the event this does not provide satisfaction for both parties, then *arbitration* should be the next step. This will require attorneys but not a judge or a hearing in court. If at all possible, arbitration should be binding. This means that you agree in advance to the decisions reached in arbitration without further recourse. AAA (American Arbitration Association) can be reached at 212-484-4000. BBB (Better Business Bureau) can be reached at 800-537-4600.

❀ In choosing designers, architects, and contractors, find people who are flexible and creative. If their pricing seems higher than it should be, then it may be because they are not as experienced as others in dealing with your desired style. They may not be as comfortable as they should be with the plans, and therefore may add to the price to cover themselves.

❀ The best way to control the price is to control the amount of detail in the design. The more intricate the work, the more costly the job. Understanding this will go a long way in helping keep cost down to a reasonable number while still allowing reasonable quality.

❀ Understand that there are contractors, and then there are *contractors*. Some have offices and a staff, while others work out of the back of their trunk! Be careful.

❀ There is no such thing as "perfect." Be reasonable in your expectations.

❀ Tips for handling problems with the contractor's work:

1. Call first and explain the problem. Do not just complain to your family and friends. I have often told clients that I can't fix what I don't know!
2. Explain exactly how you want the problem resolved.
3. If you do not get a response after a reasonable period of time, call the Better Business Bureau.
4. If the BBB has not been able to get the problem resolved, then it is time to try mediation or arbitration.
5. If this has not been agreed to, then your only option is the courts. Call your attorney!

❀ Permits are issued on a first-come, first-served basis. As a result, it is not unusual to have to wait four to eight weeks for a permit. But you can begin some prework while waiting. For example you could remove old cabinets from your kitchen and begin prepping the walls. However, be aware that if you actually begin construction before receiving your permit, you will be fined as much as $500. Worse yet, they may make you rip it all out and start again!

❀ Any remodeling project will require some of the following permits: building permit, historic permit, zoning approvals, coastal and environmental permits, and septic system approval permit. Be sure you know what you need.

❀ As part of your resale-value checklist, take into consideration the placement and number of bedrooms that makes the most sense. According to Robert Irwin, author of *Buy Right, Sell High,* most people want three or four bedrooms. *Master bedroom:* facing the street is less desirable because of the noise. *Best:* master bedroom isolated from entertainment areas in home. *Kitchen:* should be large and well lit. *Best:* an island with counter space. *Bathrooms:* at least two, preferably three. *Garage:* most people prefer a two-car minimum. *Heating systems:* Economics is the biggest concern. Gas, oil, and high-tech furnaces are much less costly than electric heat.

❀ To obtain loan interest information, visit Web site www.bank-rate. com. It compares loan rates for over 2,500 banks.

❀ If your home is listed on the preservation list as historic, you will need a special permit and will be required to work within the government guidelines for restoration. Start by contacting your state historic preservation officer (SHPO) to help you find preservationists and contractors who specialize in such work. The Advisory Council on Historic Preservation has a Web site you can consult: www.achp.gov/shpo-thpo.html. There are also architects who specialize in historic preservation. To get a list of members with this specialty contact the AIA at 202-626-7300, or visit their Web site: www.aiaonline.com.

❀ If you live near a body of water or wetlands, you may need a special permit. Check with local and federal regulators. I had an unfortunate experience with wetlands that had been filled without a permit many years before I owned the property. As a result, I spent a lot of money and wasted a lot of time trying to resolve this problem.

2

PLANNING FOR AN IMPROVEMENT

Your Lifestyle, Your Home

Without a doubt, the planning stage is the most important and exciting aspect of any improvement/renovation project. As I said in the first chapter, only you can decide how involved you want to be. It is possible to hire a professional to take charge and design what they think is the perfect solution for your home. But recognize that because every member of the family will be affected by this project, this should be a family project.

Whether you decide to hire a professional or not, I recommend that you begin the planning process on your own. Begin by writing down your *expectations*.

❀ What exactly are the problems that need to be solved?

❀ What additional functions would you like to incorporate in your spaces?

❀ Do the traffic patterns need to be rearranged, or do you just need more room, such as a family room or an additional bedroom?

❀ Is the house dark because of the lack of natural light? If so, how will you arrange the windows for your new plans, and have you given consideration to its effect (fading of fabrics, glare, etc.)?

❀ Do you like the overall look of the exterior design of your home? This is important if you are planning an addition, as it may be possible to change the overall appearance of the home at the same time.

The planning stage is the time to let your imagination go wild. It is an opportunity to explore moving, adding, or eliminating walls and windows. It is also the perfect time to think about the kind of additional storage space that would be most beneficial, like closets, cupboards, shelves, and niches.

Consider how formal or informal your life is and compare that to the overall style of your home. How can the two be brought closer together? That should be the goal: matching lifestyle with function and design. For example, would a mudroom (a place to put coats and shoes) help to resolve some of the cleaning issues in your home?

❀ Do you have any museum rooms? (For example, a living or dining room that almost never gets used?) This should be considered in the plan.

The idea is to get you thinking with a larger picture in mind. Often we can become so myopic in our scope that we focus only on one specific area, when in fact, much more can be accomplished in the process.

One of the biggest areas of consideration is your family's lifestyle. Is your home conducive to your life now? Would it make more sense to have a more *open* floor plan? Or is that the problem—not enough private space? Are your children young and wanting to be near Mommy as all times? Or are your children older and now seem *too* near? In planning the layout, it is important to realize which rooms are functioning well together and which *should* be functioning well together. A good example of this is the relationship between the kitchen and the dining room. The goal should be to allow a feeling of inclusiveness with the dining room while still providing Mom some space for working in the kitchen.

Deciding on the *scope* of the renovation is also critical. Too often, the focus is kept on one specific room without regard for how it will affect the rest of the house. Many people neglect to see how changing one space or room will negatively impact the appearance of the rest of the home. Imagine a new, up-to-date kitchen—full of natural daylight, light and airy, done in current colors, furnished with new appliances, and well organized. Now imagine how the rest of the house looks in comparison—dark, dingy, out-of-date, inconsistent in style, and depressing! This is not what you set out to accomplish, but it happens all the time. You should be both financially and emotionally prepared to accept how large the scope of a project *should* be in order to keep the entire look of your home consistent.

Inside Out: Style, Structure, and Design

There are two aspects to every home-improvement project: the *inside* effect and the *outside* effect. This is more apparent when you are considering an addition to your home. But it is also an important aspect to consider when updating the interior of your home. How your home looks on the outside should have some relevance to the appearance of the inside. Attempting to make an informal suburban ranch have the look of the Taj Mahal on the inside would be very disconcerting, to say the least. Recognizing the exterior's impact should also be a major consideration. Even something as simple as changing the size or style of a window can open up a whole can of worms. Window and door placements that make sense on the outside may not make sense on the inside, and vice versa. This is one of those areas that have become a real pet peeve of mine. For things to make sense in both places often means rearranging the entire plan. But during the organization stage, it only takes an eraser and some time. Why not make the effort? You may find a whole new solution you never considered. You may want to switch the dining room with the living room! The point is to create a space that flows—one that makes sense from one room to the next.

The first place to begin when considering an improvement and/or an addition to your home is at the building code office in your community. You need to be aware of any *setback* requirements. In other words, how far must your home be from your boundary lines? Each community has its own whole list of dos and don'ts that must be considered. They determine such things as the permitted size and shape of a structure and how far from the road or neighbor you may extend. In addition they may include such things as safety standards, exits, fire protection, ventilation, energy conservation (for example, the number of gallons of water your toilet may use!), and even lighting. These considerations will affect any renovation project. You definitely want to start out on the right foot with the building inspector. He can make your life very difficult if you do not comply with all of his rules! Knowing and understanding the "building envelope" (parameters) will go a long way in saving you time and money. By setting limitations it will also give you a basic space in which to work.

Currently, an international building code (IBC) is under formula-

BEFORE ADDITION PICTURE

AFTER ADDITION PICTURE

DETAIL OF ADDITION

tion and will be introduced in the year 2000. It will consolidate existing building codes in the United States and will be used throughout the nation, with applications in other countries to follow in the near future. In addition, a new model code is under development, called the international residential code (IRC). It will govern the construction and finishing of residences and replace the CABO one- and two-family dwelling code. This will have a major impact on what you may or may not do within your own home. For example, they are proposing a permit requirement for anyone specifying movable partitions, cases, counters, or furnishings under six feet high. The IRC will more strictly govern the use of materials for residential projects than current codes do. Personally, I am trying not to overreact, since this is still in the development stages, but it does concern me.

Be aware that in the future there will definitely be more government regulations to consider when planning residential renovation.

The Addition

When considering an addition, I highly recommend that you drive around and look at as many additions in your community as you can find. All too often, an addition may completely ruin the architectural balance of a home. I cringe whenever I see a home that has been ruined as a result of poor planning. The worst part is that in most cases it was not a matter of cost. An addition that is too contemporary or formal on the back of a traditional colonial home makes absolutely no sense to me. Two homes in my neighborhood come to mind: the first one is in a nice residential neighborhood. The home originally was a standard two-story colonial. They added a two-and-a-half-story addition to the back of the home without altering the original structure. The new addition includes a balcony with marble pediments and railings. The center has a fountain, with an open-courtyard entertaining room. As you walk through the house from the front door to this new addition, you feel as if you have just gone through a time-warp tunnel and entered the palace of Versailles. It is a very weird feeling, I assure you. The second home also started out as a two-story colonial. Several additions have been added, each with its own style and architecture. This particular home has been for sale for three years. Because of all the money that has been invested in these different additions, the home is priced way out of its market. Furthermore, it will take a very special person to find this home appealing. It no longer functions well as a whole, and you nearly need a road map to find your way from one space to the next. What makes both these situations so sad is that a lot of money was spent to create them. Who is at fault? I don't know. I was not involved in these renovations. But I cannot imagine the homeowners paying for them if it did not meet their expectations. How would I have handled a client who desired such work? I know I would have done as much as I could to enlighten and educate with regard to style, function, and reasonability. Not knowing the owners, I have no idea who was ultimately in charge of these projects. My goal here is just to make you aware so you can avoid making similar mistakes.

Here are some tips that may help you avoid monstrous mistakes:

❀ Match materials and roof styles to your existing home.

❀ If you are determined to depart from the original home style, then incorporate some of the original design elements or materials of the old.

❀ If you are not confident, hire a professional for help. This can be a builder, designer, or architect. (I'll explain further in this chapter the differences and designations for each.)

❀ If obtaining matching materials is not possible, then consider changing elevations. In other words, set the new structure back or forward a foot, and use a complementary material. For example, a stucco finish will complement brick very nicely, as will vinyl siding. My own home has a combination of all three. The key is justifying the different material with the setback.

❀ Mimicking or imitating your existing structure in a smaller form is also a good way to avoid unpleasant style mixing.

Floor plans play one of the biggest rolls in creating a new addition. You may or may not decide to hire a professional to do this job. Regardless, I suggest you at least do the first rough sketches yourself. Don't worry about your drawing ability—this is not a test! The idea is to help you begin to formulate your ideas in a more concrete way. Start by measuring everything, and I do mean *everything!* Not only is it important to use precise measurements of your existing home, but you will also need to have general size information of any furniture you plan on putting in the new space. By doing the floor plan for the furniture

BEFORE EXTERIOR RENOVATION

AFTER EXTERIOR RENOVATION

arrangement at the same time, you will find the whole process easier. Most people struggle with floor plans because such plans are so one-dimensional—a bird's-eye view. In my first book, *My Name Isn't Martha, But I Can Decorate My Home,* I go into an in-depth discussion on how to learn to see a room from several different dimensions and viewpoints. Recognizing which measurements will most affect your visual perspective is very helpful. A pathway of 20 inches is fine, if it doesn't impede your eye-level plane. For example, 20 inches is sufficient space for walking between an ottoman and a sofa. However, it is not necessarily an acceptable pathway between a wall and a tall cabinet because your visual plane is blocked at eye level, giving the appearance of very limited space.

The best scale to draw a floor plan in is $1/4$-inch scale. This means that each foot will be represented by a $1/4$-inch line. You can purchase graph paper and premeasured furniture templates at any office supply store to fit this $1/4$-inch scale method. In these early stages of planning, you should not lock yourself into any specific placement. Try as many different floor plan arrangements as are feasible. Unfortunately, this is where most people get stuck—they become determined to put a specific element in one specific spot, and then lose all potential for something better. Be flexible—you are not carving this in stone.

Initially, don't be too concerned with determining exact proportions of the new space. The dimensions will ultimately be dictated by the furniture, architectural elements (windows, fireplace, etc.), and building limitations. Do consider the building code size restrictions for any additional needs such as a deck or patio, and even landscaping. My backyard is part of a very large retention basin in our development. It runs through several yards on my side of the neighborhood. This means there are a lot of restrictions with regard to building. I could not, for example, even build a deck or patio beyond a certain point because it would restrict the flow of water through the basin. I am also restricted in planting and landscaping. Root systems and reshaping the topography of my yard could affect the flow of water through the basin.

Any landscape planning should be carefully considered during this phase. I knew before I began my new porch that I also wanted a patio. By working with a landscape architect simultaneously, I was able to create an interesting circular patio that may not have been possible if I had waited until the porch was finished. This was also included in the overall price. By being aware of all the costs, I could stage the project over two years, making it much more palatable for me to handle financially. I actually had the patio built the summer before the porch.

Once you have begun to establish a basic size and shape that accommodates your desired elements, it is time to make the space more creative architecturally. Think about the different kinds of windows and the effects they could have. This may also be a good time to begin getting a professional involved in the process. The reasons are simple—engineering the structure for both stability and cost effectiveness is critical to the design. Codes and the complexity of the renovation will also affect what you can and cannot do. For most projects, you will need drawings to present to the building code office for review. Unless you are a qualified draftsperson, then you will need to hire someone anyway. For my most recent addition, I used the builder's designer. He created the basic drawings on a CAD (computer-aided design) program. They reflect all the details needed to determine the feasibility of the project. In addition, they were all labeled and coded to industry standards. This makes them understandable to everyone working on the project. If you are contracting this work out, most builders will provide basic drawings as part of the bidding process, free of charge. If you decide to draw your own plans, you may risk being misunderstood. This can cost you time and money. CAD systems are available to the public, but unless you plan on using a program like this a lot, I have found they take far too long to master. And you are still faced with meeting codes, regulations, and so forth. You also may need to present a plot plan showing the extension of the new addition in relationship to all restricted boundaries.

Another consideration when planning an addition or renovation is *exposure*. Exposure refers to the direction your room is facing, that is, north, south, east, or west. You may envision a room of windows that allow the sun to beam through and fill the space with warmth and light in the winter. However, if your exposure is northern, this will not happen. You must also consider the position of the sun in the sky for each season, and how this will affect heat and lighting in your new space. The sun is lower in the sky in winter. If you can take advantage of this then the result will be very energy efficient. But you must then remember to add eaves (overhangs) to limit the sunlight in your room in the summer—otherwise you could get real hot under the collar, come the warmer months!

Professional Who's Who

In almost every discussion on hiring a professional, there seems to be one continuing concern or reluctance: the fear of having the professional's opinions and ideas forced on the client. Yes, money is also part of that discussion. But I think most people are more afraid of having to satisfy a professional's *ego* rather than of having their own needs met.

The fact is that qualified, competent professionals, whether they be architects, designers, or even builders, can often do more in helping you turn vague ideas into real, creative solutions that make sense from all aspects. Of course, there will always be those few who ruin the reputation of the entire bunch. That's the reason for checking references. By doing your homework and exploring your own ideas and desires you will be better prepared to hire the right professional for the job.

So who is the right professional? That will depend on both the extent of the project and your own desires and abilities. Here are the basic descriptions of who's who.

American Institute of Architects (AIA)—These words (or initials) designate a licensed architect who has a bachelor's or master's degree in architecture from an accredited university, and has completed an internship under a licensed architect, usually working as a draftsman. Architects are registered with the state, having passed a registration exam that lasts several days. They are tested annually and are required to complete thirty-six hours of continuing education each year.

Warning. Not all who describe themselves as being an "architect" really are! Some who use the term "architect" may have graduated from a four-year program or even a graduate program, but they may have little or no work experience. In this case, they probably have not taken the registration exam and are therefore not officially registered with the state. Nor are they required to take continuing education courses.

However, they can be a cost-saving way of getting good ideas, as long as you are aware of the risks. They may or may not be knowledgeable in areas of technical excellence or have had much actual experience working in the field. This may make them less creative in terms of possible solutions to your needs.

An unregistered "house designer" architect has done a lot of work here in Lancaster, Pennsylvania. Unfortunately, after many years of having a good reputation, he had a structural problem with a home he had built. A suit was filed, and he is now no longer able to practice.

What to Expect. First, expect the architect to show you photographs of previous work. Specifically ask to see projects similar to your expectations. At your first meeting, be prepared with your own rough sketches, your expectation list, and any photos you may have found from other sources (architectural magazines, books, etc.) that appeal to you. Be prepared to let him know all your family secrets (well . . . almost all!). It is necessary for the architect to know how you function in order to help you. And definitely discuss your financial budget. It is impossible to begin designing without an idea of how much you are willing to spend.

Most architect's fees are based on a percentage of the total project cost. Usually this is ten to fifteen percent. Be sure you know exactly what will be provided for this fee. Most often, a fee of this nature includes the following:

❀ *Design analysis and development.* This is the first real discussion on what you expect and the analysis of whether or not it is feasible.

❀ *Schematic plans.* These are the preliminary sketches and a rough cost of the project. This is the time to really open up and express yourself—or forever hold your peace. It is an opportunity to see if and how all your dreams can be realized.

❀ *Design development.* This is when the details of the design are added. This usually includes basic structure, engineering, and preliminary landscaping. If your project includes a kitchen or bathroom, it may also be the time to get a kitchen or bathroom designer involved, because they are specialists at planning those types of spaces. At this stage, you have an opportunity to make *minor* changes.

❀ *Construction documents.* No turning back—your plans are being finalized! This is when the actual working drawings and plans will be produced. They will include all the specifications required to finish the project. By now, you should have chosen your flooring, wall covering, and/or paint (one, two, or three coats?), your cabinetry, etc. In other words, *everything* needed to finish the project.

❀ *Bidding.* Here you have a choice—you may or may not decide to keep the architect involved in the bidding process. When I last worked with an architect, it was a fairly large project. I wanted to get several bids, and as a result, I felt it necessary to keep the architect involved. It made it easier to be sure everyone was bidding on exactly the same things—in other words, apples to apples, and not apples to oranges. I do recommend that all bidders spend time "on site" to be sure there is no guessing about what's what. Also remember, price is not everything. The contractor you choose should be someone you feel you can truly work and communicate with on a mutually beneficial level.

❀ *Contraction management.* The architect should hold regular meetings to address any questions or changes that may arise. In addition, he can monitor progress and approve the payment schedule for the stages of progress. You may or may not decide to have the architect involved in this process. Oftentimes, a general contractor will do just fine. If you are working with a loan lender, then often the contractor can present the lending institution with inspection certificates along with your payment at each stage of completion. Most architects will be willing to offer limited services in this area. If a question or a need arises, you could also pay by the hour.

American Society of Designers (ASID)—Designates those who have a bachelor's or master's degree in interior design and at least two years' experience. They have also passed a qualifying exam. An *Allied* member is one who has been qualified through education or experience for membership but has not passed the exam.

As with architects, there is a lot of confusion as to who is who. Most professional designers (that is, members of ASID) use the title designer. Most others (who are not ASID) use the title decorator. But the only sure way to know their qualifications is to ask.

Warning. Many states now require licensing for designers and many others are considering it. The qualifications and entrance examination are different for each state. There is ongoing lobbying to continue with licensing across the nation, but it is a long and costly process.

What to Expect. Designers specialize in *interiors.* They will help with conception and layout of interior spaces, helping to determine the best placement of doors, windows, and electrical outlets, as well as developing the basic lighting plan. They will also help choose finish materials such as paint, stain, flooring, fixtures for bathrooms, and surface materials—for example, kitchen counters. In addition, they can help select furnishings, window treatments, fabrics, and decorative and architectural elements. As a designer, I know all the options that are available for your interior finishes. I can help you make good choices that function effectively while meeting your budget. The creative use of materials is what makes your home unique.

Designers also help you avoid costly mistakes by assisting with product selection and taking detailed measurements. Designing the interior of closets, bookcases, and other storage areas is another area where they can be very helpful. Sometimes, a few inches here or there can make a world of difference. A good example of this is in the placement of windows. Too often there is not enough space left between the corner and the window casing. This makes it difficult to design and install a window treatment that best complements the room. A few inches here can be all that is necessary.

 Many times, I have custom-designed furniture and cabinetry to specifically meet the needs of an individual client. Planning for such items early in the development stage was critical to the success of the project. In one particular instance, I noted when visiting the construction site that the ceiling in the loft area was too low, thus making the space impractical. On my advice, the roof was raised to accommodate a space that then functioned usefully.

A designer's goal is to work closely with clients in order to develop a style or central theme for their home's new look. Professional designers are one of the best resources for knowing where to find products for every aspect of the job. They can introduce you to the newest design product ideas and conceptualize the best arrangement for your living space.

Most designers have the training and expertise to manage all the details of finishing a project. Many designers also have an architectural degree as well; estimates indicate that nearly half of all designers have such a degree.

National Kitchen and Bath Association (NKBA) CKD & CBD—This designates a person as a certified kitchen designer (CKD) or a certified bathroom designer (CBD). Talk about specialization! They too must pass a design exam. In addition, they must have at least seven years' experience working in the industry. At least two of these seven must be in full-time kitchen or bathroom design.

Today's kitchen has become the most sophisticated area of the home. As a result, the technical, functional, and visual aspects require a specialized planning process. Most architects are not qualified to do this work.

Most kitchen/bath designers work for one specific cabinetry company. This means that who you choose is often dependent on which cabinet company you choose. I suggest getting plans and pricing from at least two within your specific budget range.

What to Expect. Obviously this will, to some degree, depend on the scope and budget of your project. But for the most part, a kitchen/bath specialist will be responsible for the following:

❀ Specifying and coordinating all aspects of the project. This will include designing the space, drafting working drawings, and doing all the necessary specifications for cabinetry, including interiors and installation.

❀ Specifying countertops, appliances (this can be a nightmare), and making provisions for necessary measurements and ventilation needs.

❀ Ordering all materials and scheduling the installation process.

❀ The management, and often the hiring, of all who are needed to successfully complete this project. This means: electricians, plumbers, carpenters, and countertop professionals. The intricacy and detail to the tiniest measurement can make all the difference about ensuring a successful outcome.

National Association of the Remodeling Industry (NARI)—This is the designation for all those who have been in the remodeling business at least five years. In addition, they must pass an eight-hour exam and take continuing education courses. **CR**—Designates a certified remodeler. **CRS**—Designates a certified remodeler who *specializes* or focuses on one area of the industry. CRSs also take the business management and remodeling sections of the exam in their specific field.

Remodelers Council of the National Association of Home Builders (NAHB)—This is the designation for those who have either owned or managed a business for five years or more. **CGR**—Designates certified graduate remodelers. They also have education and experience in remodeling. CGRs must recertify every three years by taking continuing education courses.

General contractor—The builder of your project usually assumes this role and is the person responsible for all facets of the project, including scheduling subcontractors and making sure all necessary supplies are where they need to be, when they are needed.

Checklist for Budgeting

I've included the following checklist for planning your budget. Hopefully, it will provide you a way to see clearly all of the costs that can be involved in renovating your home. This can be helpful in working with possible contractors for establishing a guideline for budgeting. I think it is also helpful to see where the money is allocated. You may find that there are areas you can do yourself and thereby save some money and maybe a little time. I've also included some rough estimates that can be used to give you some indication of what things should cost.

- Design and/or building plans
- Permits—building, plumbing, and electrical
- Demolition and disposal
- Carpentry and drawing
- Windows, doors, cabinetry, installation, and finish carpentry
- Countertops and backsplashes
- Appliances
- Lighting
- Mechanical—plumbing, electrical, heating/air-conditioning, venting
- Flooring and subflooring
- Wall covering—paint and/or wallpaper

Here are some average costs for typical aspects of your budget:

FLOORING (BASED ON A 15-BY-20-FOOT ROOM)

New subfloor (plywood overlay): $450 to $600
New resilient floor: $15 to $45/square yard
New ceramic tile or wood floor: $30 to $70/square yard

PLUMBING

Replacing a sink with a new one in the same spot: $150 to $250
Running a plumbing line to a new island: $300 to $500
Moving a gas line: $200 to $300
Adding a second or double-bowl sink: $200 to $300

ELECTRICAL WORK

Outlets: $50 to $75 each
New fixture: $30 to $100
Installing new light fixture: $140 to $175 (including new switch)

DEMOLITION

Removing a non–load-bearing wall (15 by 8): $300 to $1,800. This will depend on the necessity of relocating any plumbing or electrical lines.

Moving a load-bearing wall: $700 to $5,000. This will depend on the necessity of reinforcement, repairs, and any plumbing and electrical work.

Moving an interior door: $800 to $1,200

Closing up an exterior door (including siding): $400 to $800

Moving an exterior door: $800 to $1,200. This will depend on whether you need to cut through brick or siding.

Adding a pocket door: $300 to $600

Closing an existing doorway: $400 to $800

Replacing a window: $150 to $250

Installing a skylight: $275 to $800

Moving a window: $800 to $1,500

Replacing a window with a greenhouse window: $200 to $1,200

Words of Wisdom

❀ Regardless of the kind of renovation project you are doing, you will need a building permit and the project will need to be inspected at regular intervals. For example, the electrical and plumbing work will need to be inspected when the rough work is complete before you can close up the walls.

❀ Although private homes are exempt from the ADA (Americans with Disabilities Act), you may want to consider some of their rules for your own needs. I did in my own home, with a one-floor living plan. As I age, I want to be comfortable, and therefore I eliminated the need to climb stairs. Here are some other specific suggestions:

- ❀ Wheelchair access requires doors to be 36 inches wide.
- ❀ Kitchen countertops should be 32 to 34 inches from the floor for a wheelchair.
- ❀ Use low-pile carpet glued directly to the floor.
- ❀ Hallways should be 36 to 48 inches wide.
- ❀ Make the bathroom safe by planning for grab bars and allowing adequate space in the shower for a chair.

❀ Is your house structurally sound? Try jumping up and down in the center of your rooms—check for springy floor while listening for strange sounds.

❀ Benjamin Moore Paints offers Website how-to information for a variety of projects such as painting your vinyl floor to look like faux terra-cotta tile. Web site: www.benjaminmoore.com.

❀ If you are not the first owner of this home, be sure that you specifically check the deed for your individual property and any unusual boundary and/or code restrictions. They should be included in your deed but often are overlooked several owners later.

❀ Here are the numbers for some trade resources:

American Institute of Architects (AIA)—800-242-9930
American Society of Designers (ASID)—800-775-ASID. Web site: www.Interiors.org
National Association of Home Builders—202-822-0216
National Association of the Remodeling Industry—800-440-6274

National Association of Plumbing, Heating and Cooling Contractors—800-533-7694

American Arbitration Association—800-778-7879. Web site: www.rahul.net/designlinc/desingo.htm. This site provides advice to "universalize" every part of your home. It shows how children, the elderly, and those who have disabilities can all be included and benefit from universal design plans.

❀ Always get more than one bid—three to five is best. Choose based on reputation and your ability to communicate rather than on price alone.

❀ In selecting a professional, it is important that you feel comfortable. Choose designs that suit your individual taste and budget. Be sure the professionals you choose have the training and expertise to manage all the details of the project, from foundation to finish.

❀ Even with the change in capital gains rules, home-improvement records are still important. Your first $500,000 in profits from selling your home are nontaxable for couples, $250,000 for singles—but cost *basis* of the home must be shown when filing tax returns. If the gain is above the excluded amount, extra *basis* will save you taxes. Home *improvements* increase your basis if they add value to the house, extend its useful life, or adapt it for new uses. Home *repairs* generally do not add to the basis, unless they are part of an extensive home improvement.

❀ There is a new mortgage available. It's called the HomeStyle mortgage from Fannie Mae. It is based on a home's appraised value after renovations are finished—not its value at the time of application. This allows you to borrow enough money to help finance major repair work. Many lenders offer the HomeStyle mortgage at interest rates that are comparable to standard mortgages. For more information contact Fannie Mae—800-732-6643.

FLOORING AND
LIGHTING—TWO
ESSENTIALS

Flooring

The biggest mistake most people make when choosing flooring for their home is to make choices for each room or space individually, instead of considering a cohesive plan for the entire house. Just like any other aspect of home interiors, you need to design a plan for your flooring. Floor material choices should be coordinated not only by color but by overall style as well. Your home should have a consistent personality that is representative of your lifestyle. Some people are more formal, while others prefer a casual style. Obviously, some rooms can be more formal than others, but it doesn't make sense to change from something as formal as Gothic in your living room, to American country in your dining room. It would be like traveling in a time machine. Transitions need to be much subtler, allowing your eye to actually believe these things belong together. How do you accomplish this? By planning ahead. Your main living space (often your first floor) should have no more than five different choices of flooring materials. These five choices also should all be related in color to present a variation of compatible textures. A good example of this would be to use two different colors of similar tile (one in the kitchen and one in the foyer), a hardwood for the living and dining rooms with a pile carpet or oriental carpets as area rugs over the hardwood, and a loop (textured) carpet for the family room. You could then choose more informal area rugs to use in the kitchen to convey a less formal atmosphere.

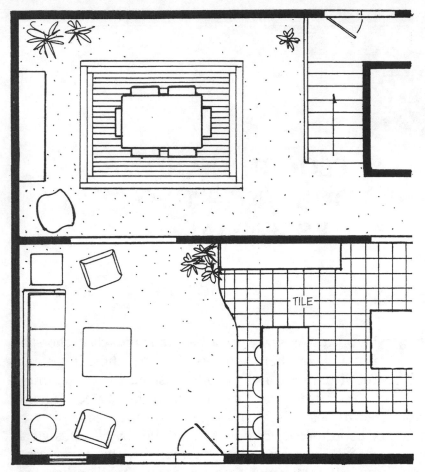

TILE

CARPET WITH WOOD FLOOR INLAY AND WITHOUT

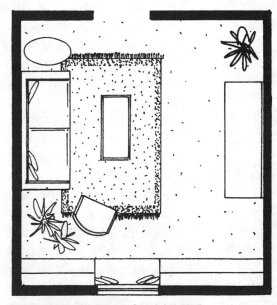

ORIENTAL CARPET ON CARPET

ORIENTAL CARPET ON WOOD FLOOR

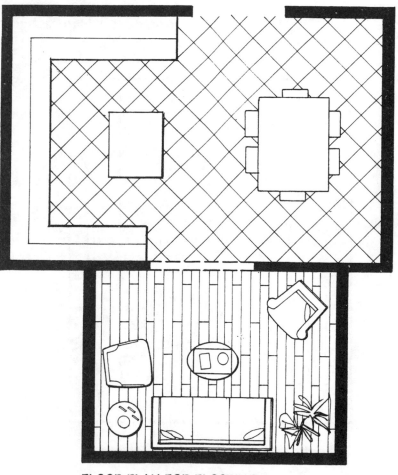

FLOOR PLAN FOR FLOORING

Determining where one material stops and another begins is very important to the overall scheme and appearance of your home. It can make the difference between a space seeming larger or smaller. It can also affect the placement of furniture. In my own home, the archway between the kitchen and the great room offered some flexibility, because the wall between them was the fireplace wall. Since the fireplace wall is deeper (thicker) than a standard wall, I could change floors almost anywhere within that dimension. By laying out my furniture plan first, I was able to determine how much room I would need to place my favorite bench. I wanted the bench to be a part of the kitchen, not the great room, therefore, I needed to extend the resilient kitchen flooring into the wall dimension enough to accommodate the bench. The first consideration should always be the architectural boundaries. Such things as walls, cabinetry, and archways are good choices for determining such changes. Always be sure to consider your furniture plan as well. If you have symmetrical spaces, such as a living room and dining room off either side of your entryway, then you should consider using the same flooring materials in each to create a balanced look.

MY FLOOR TRANSITION FROM LIVING ROOM TO KITCHEN

It is also important to give consideration to the direction of such things as tile and wood planks. Often I recommend placing tile on a diagonal. This creates the illusion of more space and eliminates the confusion of aligning wood planks evenly with the tile. I also suggest that wood planks be directed from the front to the back of the house. This too will give the impression of more space. The direction of the wood planks will actually lead the eye to follow this direction. Take this into consideration and give the eye something to see at the end of this space. If it's a wall, add something interesting to it, like artwork or a mirror.

The second-floor color plan should be similar to the first. One of the most often asked questions is, Can I use a different color flooring in each bedroom? The answer is yes, however, you must be sure that the color schemes you choose are compatible with the main color scheme of the house. The second-floor hallway should be in the same main color as the first floor—although it can be a different texture. Each adjoining room should follow a color palette that is compatible with the hallway. For example, if you have chosen spruce green as your main color, then beige, copper, and deeper or lighter shades of spruce would work, as would a coral and spruce combination. The key is to tie in the spruce green in some way in each adjoining room.

To help accomplish a cohesive flooring plan, start with a comprehensive sketch of your home. It's not important that your measurements be accurate—the idea is to establish a "big picture" view of your spaces. Note the natural architectural breaking points—such as doorways and stairs. Then consider the following needs:

❀ The individual functions and amount of traffic of each space.
❀ In which rooms are spilling or staining accidents most likely to occur?
❀ Be sure to recognize where water is an issue—kitchen, bath, laundry, and basement.
❀ Analyze your budget costs and determine how many and what types of materials you can afford.
❀ Consider the need for thresholds to even out the differing heights when transitioning from one flooring to the next. How will this affect subflooring?
❀ Maintenance—be sure you are realistic about how difficult a material will be to keep clean.

A Note about Kitchen Floors. Because your kitchen is such a highly trafficked space, the choice about the flooring you make is important not only for the immediate aesthetics and style, but also in determining how well your kitchen will look after years of use. You should choose a floor for its appearance, performance, maintenance requirements, and its cost. Most floor products are now universally used throughout the home. In today's market, most products are available in a variety of prices, making it possible to choose almost any type of floor you desire. For example, in the past resilient (vinyl) flooring was chosen because of its affordability. Certainly, there are still some very reasonably priced resilient floors. However, there are also some very expensive resilient floors. Which means that one reason for choosing a resilient floor is for its performance attributes and not its economic advantage.

When considering a new floor for your kitchen, the first thing you need to do is examine your existing floor. The reason for this examination is to determine if the old floor will have any long-term effect on your new floor's appearance or performance. If your old floor is vinyl, I suggest looking at it from an angle, with light shining on it. This will allow you to see any imperfections such as ridges, bumps, and valleys. If it appears to have an even ridge pattern, it means your vinyl floor was probably installed directly over a subflooring. However, if you see a typical vinyl floor pattern (such as flowers, checks, and/or a lot of texture) it was probably installed over an existing vinyl floor. This is not uncommon or necessarily wrong. I would not, however, recommend adding yet another layer without first installing a new underlayment, particularly if you are planning on a new ceramic tile or stone floor.

You can, of course, tear out the old floor. But I caution you to be sure that the old floor is not asbestos. If it is, *do not* remove it. As long as you don't disturb asbestos, it's safe. Just install a new underlayment on top of it and proceed from there.

If upon examination of your old floor you discover it is uneven, you can use a liquid leveling agent to even out the peaks and valleys. Depending on how uneven your particular floor is, you may want to add a 3/4-inch plywood subfloor as well. This will also work if you have a combination of floor surfaces to work with. This is often true when changing and/or adding to the kitchen floor pattern. If the change of levels is too great, it may be worth considering a way to incorporate

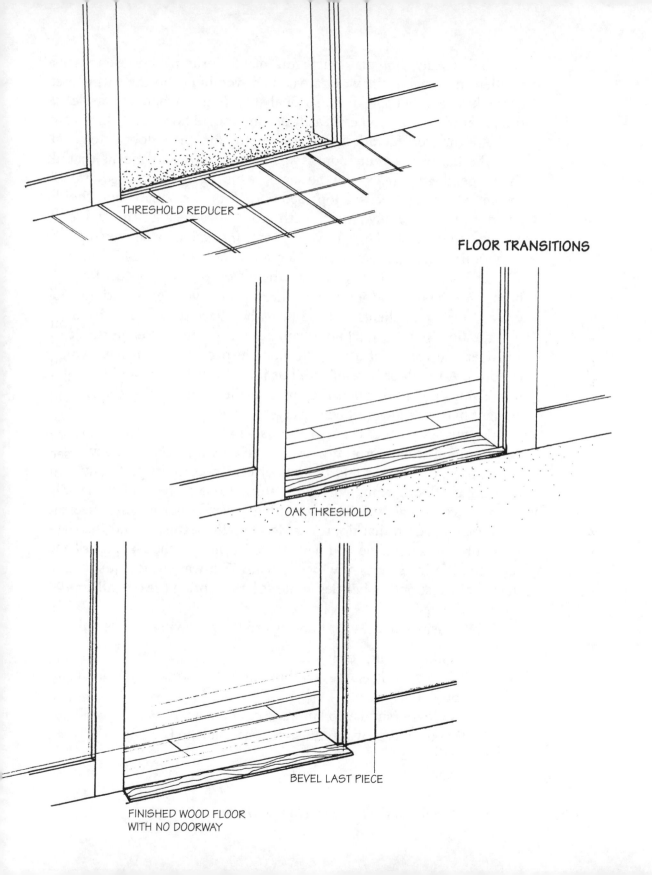

THRESHOLD REDUCER

FLOOR TRANSITIONS

OAK THRESHOLD

BEVEL LAST PIECE

FINISHED WOOD FLOOR
WITH NO DOORWAY

two different flooring materials. You may for example choose to use a resilient material in the work area and a wood floor in the eating area of the kitchen. By using a transitional strip (either wood or marble) at the point of change, you can easily accommodate floor level changes.

Another consideration before shopping for a new floor is to check out whether or not your room is square (plumb). This means that all the corner angles are exactly the same, resulting in a perfect square (or rectangle), as opposed to a lopsided one. If it's not square, you should avoid geometric or any pattern with grout lines because this will only accentuate the irregularity of your room. Instead, choose an irregular, or nondirectional pattern such as a granite look, or even a leaf pattern.

The best advice I can give you when choosing any product for your home is to be realistic about your lifestyle, and your expectations. For years, I have had clients tell me how they have dreamed of having a real tile floor in their kitchen. If you are one of those dreamers, please consider the reality of a real tile floor, especially if you have young children. A tile floor is beautiful, but it is also very hard and unforgiving. Things do break when they meet a tile floor with force. This includes children and your favorite crystal vase.

Moving on to the rest of the house, I always dreamed of having a white ceramic tile floor in my foyer. While living in New Jersey at the shore, I decided this would be the perfect opportunity to finally have that dream floor. Unfortunately, it turned out to be a nightmare! It showed every single speck of everything—even airborne pollen seemed to be magnified on that floor. I came to really despise it and the work it required to keep it clean. I have since learned to choose *speckled* patterns in differing shades of white instead. That way the dirt just blends in with the rest of the speckles while still presenting the overall appearance of a *white* floor.

Your basic materials for flooring are the following:

- ❀ Resilient—vinyl or linoleum
- ❀ Tile—ceramic, stone, marble, slate, or pavers (brick and terra-cotta)
- ❀ Laminate (similar to laminate countertops)
- ❀ Wood—old, new, prefinished, or unfinished
- ❀ Carpeting
- ❀ Concrete

I covered all of these floor styles in great detail in my first book, *My Name Isn't Martha, But I Can Decorate My Home.* Here I will give you the basics to help you narrow down your search.

Resilient flooring refers to both vinyl and linoleum. *Vinyl* flooring is still one of the most popular choices in the kitchen. It offers the widest variety of price, performance, and style. I recommend buying the best you can afford because there are valuable differences in quality and performance from the low end to the higher end of the spectrum.

Vinyl is available in two basic forms, inlaid (the color and pattern are uniform throughout) and rotogravure (the design is printed on the surface). Many consider inlaid to be the most durable. However, I believe that the wear layer (finish coat) is just as important to a floor's durability. High-end floors have a urethane wear layer. This is the best stain resister and shine retainer available for vinyl flooring. Midrange vinyl floors may also have a urethane wear layer, but it is usually thinner than the higher priced floor. The best way to know how well to expect a floor to perform is by checking the length of the warranty. If it's a three-year warranty versus a ten-year warranty, it's easy to figure out which one will last.

Over the past several years, the price of vinyl flooring has really increased. In an attempt to make available more choices at a lower price point, Armstrong Flooring has reworked most of its line. They have actually taken their lower-priced flooring and added the higher-priced finish to it. This does not mean that it is equal to the high-priced floor. It does, however, provide a better product for less money. You still get what you pay for.

Maintenance is really easy with the new vinyl floors as long as you follow the manufacturers' recommendations and use the products specifically designed for your choice of floor. Installation will depend upon the specific floor. Some only require glue around the perimeter, making them a good choice if you are a DIY person. Price range (for a 12-by-15-foot room): $200 (budget flooring): $900 (high grade).

Linoleum is an all-natural product made from either linseed oil, pine resin, or wood flour. It has been out of style and nearly out of existence since the 1950s except for some production in Europe. The basic two style choices were and are either a solid or a marbleized look with a matte finish (price range: around $4/square foot).

Tile is available in a vast majority of styles and materials. The word *tile* is actually more representative of an installation technique than of a product or material. Today, even marble and carpeting are available in a tile format. Within this array of choices, you will also find an equally vast number of price ranges ($10 to $20/square foot).

Often people complain that tile is cold. In fact, it can be cool. On the other hand, it can also act as a passive solar heat unit. If you use a dark color tile in a space where it can absorb sunlight/heat during the day, it will dissipate this heat into your room when the sun goes down. It's a great way to gain heat in the winter. Just be sure to shade the sunlight out in the summer. The sun is higher in the sky in the summer, making it easy to shade out with an awning. If this is not an option for you, and you live in a cold climate, then install radiant in-floor heating units. This will give you the option of keeping your toes warm by heating the tile from beneath the floor. It's fairly easy to install; you just need to plan ahead so you can lay the required wiring before installing the tile.

Ceramic and *quarry* tiles are available at the same price as a vinyl floor. Obviously you will have fewer choices of style and color at the lower price points, but you shouldn't have to compromise quality. Most floor tiles are 1/4- to 1/2-inch thick. Do not get floor tiles confused with wall tiles. They are not interchangeable.

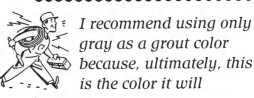

I recommend using only gray as a grout color because, ultimately, this is the color it will become!

I prefer larger-size floor tiles—it creates the illusion of more space. The smallest size I recommend for a kitchen is 8 inches. I also prefer using glazed tiles (presealed with a permanent protective coating) with a textured finish to make them slip resistant. They are easy to keep clean—just a damp mop—no waxing, no work. If you choose an unsealed tile, it will need to have a sealer applied. In addition, you will have to reseal it as needed. Otherwise, you will end up with a very stained floor. In either case, you will need to choose a grout color.

Stone, marble, and *slate* tiles are cut from natural stone. They are quarried (mostly in Italy) from marble, travertine, slate, granite, and limestone. They are usually 12 inches square or larger. Stone tiles are wonderful. They have a rich and interesting texture and color that makes them suitable for most styles of decorating. The most popular finishes are honed and tumbled. Both of these are a matte finish. Honed is smoother, while tumbled uses a technique of rotating the stone in a mixer to round off the edges, which creates an interesting texture. Over time these two finishes will develop a beautiful patina (a matte sheen). My favorite stone choice is limestone, particularly French limestone. It has a subtle elegant look with a charming Old World appeal.

Marble is beautiful, but it is also one of the more difficult products to maintain. It is usually highly polished, but this polish is not permanent. It will need to be repolished on a regular basis. It is also very slippery and near deadly when wet! It is also very porous, which means it stains easily. I do not recommend it for any heavily trafficked area. Often *travertine* and *terrazzo* are confused with marble because they are actually made from marble. Travertine is a variety of crystalline limestone (commercially known as marble). It is rarer and more expensive than other stone products. Terrazzo is marble or granite chips mixed into a cement or resin base. They are then ground into tiles and polished.

Slate has become very affordable over the past few years. The reason is that it has become a lot more available. It is now imported from several countries: China, Africa, England, India, Germany, and England. You will probably be surprised also at the variety of colors and textures now available. Recognize that it is a natural product and therefore it too needs to be sealed and properly maintained.

All stone products are usually available as a slab (versus a tile). This way you can create larger or more interesting patterns, depending on the size of slab you purchase. I do suggest shopping around if you choose to go this route. There can be a lot of variation from one slab to another.

Pavers refers to any thick tile specifically designed for floors. Originally they were developed as a commercial floor product, often seen in malls and fast-food chains. They are usually unglazed, making it necessary to maintain a sealer coat at all times. Terra-cotta tile often is categorized as a paver. However, terra-cotta is available both glazed and unglazed. And depending on the manufacturing technique, it can either be very durable or easily subject to chipping. I used glazed terra-cotta tiles (from Mexico) in my last home. It was a passive solar home, so the tiles served as part of my heating system. I chose glazed tiles for their easy maintenance. I did experience chipping on a few occasions when I dropped something heavy on them. There is no realistic way to camouflage a chip. It just becomes part of the endearing quality and style of the floor.

Laminate flooring is one of the more recent technological developments. Basically it is the kitchen countertop *pumped up!* In fact, it's so tough that it resists cigarette burn marks and staining. Regular sweeping and occasional mopping are all that are needed for maintenance. It is a laminate with a reinforced protective coating bonded to a fiberboard substrate. Most manufacturers have now added a moisture-resistant backing, which makes it suitable for the kitchen. However, I still do not recommend using them in a bathroom or laundry room. The reason is that the fiberboard base acts like a sponge and soaks up water very well. It's a "floating floor," which means it can be installed over existing floors without being nailed or glued down. Most manufacturers offer a five- to ten-year warranty. Prices range from $7 to $8 a square foot. It is available in natural-looking styles such as a wood look, as well as a variety of textures and colors. It can even be mixed and matched to create interesting borders and focal points.

Wood flooring always has been and continues to be one of our true loves. Its natural warmth is always inviting. Overall, they are easy to maintain. Vacuum or sweep and occasionally mop them when needed. With today's finishes, you shouldn't have to polish or wax them. Wood flooring is available either prefinished or unfinished. The prefinished floors have a variety of seal choices. Some are even impregnated with acrylic, making them nearly indestructible. Most prefinished wood floors have a polyurethane that should endure most situations well.

There are more varieties in the choices of wood now being produced, but oak and maple are still the most popular for more traditional rooms. For a more contemporary look, try some of the lighter shades and less grainy woods. Pine is seeing a resurgence of interest as a floor choice in the kitchen, particularly in a paler shade of stain.

Most wood floors are available in either strip flooring or parquet. Strips are either random width and length or planks (3-to-8 inch-wide boards). The use of antique or well-aged wood flooring has always been popular here in Lancaster, Pennsylvania. It is now becoming popular across the country. One manufacturer is actually advertising how many floors he can produce without felling one tree! It's a great way to get the wonderful character and charm of an older home without the hassle.

Wood flooring can be informal or extremely formal. That's what makes it so popular. Intricate combinations of pattern, wood, and color can be used to create one-of-a-kind statements.

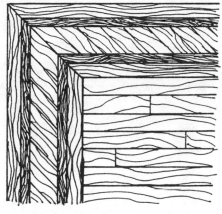

BORDERING WOOD FLOOR

PARQUET FLOORING

CHEVRON WOOD FLOOR

Carpeting offers the largest choice of styling and price available. I have become quite opinionated about carpeting. I believe that the best carpets are *commercial grade* carpets. They offer the best durability and ease of maintenance possible. But, they are also more expensive than residential grade carpeting. So I try to combine the best of both worlds by using a combination of both commercial and residential within a home. If commercial carpet is out of your budget, then choose residential carpeting that is most similar to commercial, with a lower nap, tighter weave, durable fiber, and low- or no-tracking style.

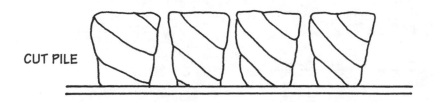

CUT PILE

COMBO CUT
PILE AND LOOP

The key aspects to choosing carpeting are: fiber or content, construction, backing, installation, and maintenance. There are basically six different types of fiber used for carpeting: nylon, olefin, polyester, wood, acrylic, and cotton. Oftentimes, you will find a combination of fibers used. Each fiber has its own characteristics and durability properties. The most common fiber is nylon. It also offers the widest range of choices for style, color, and price. Most name-brand nylon carpets are basically the same technically configured nylon fiber. Therefore you can be assured of similar quality and durability by choosing brand names such as Anso, Wear-Dated, or Stainmaster.

Be sure when purchasing carpeting to ask to see a layout plan for seaming. It is critical that the direction of the nap is consistent. In other words, the carpet should be going in the same direction at all times. Otherwise, you will have an obvious difference of color and texture.

Many people make the mistake of thinking that a thicker padding under their carpet will provide better durability. Actually, the thicker the rug pad, the more difficult it is to walk on, especially in high heels. The most important aspect about rug padding is its density. The denser the pad, the more support it will give your carpet. Some carpet manufacturers' warranties can actually be negated if you do not use the recommended pad. The industry standard is a one-half-inch-thick pad consisting of either a prime urethane or a rebond pad. The prime urethane has more air and therefore feels softer. The rebond is a denser and firmer pad.

If you are getting competitive prices from more than one retailer, be sure that you have a copy of both seam plans and that the yardage is similar in quantity. Seams should be placed in the least obvious place possible.

In my opinion, carpeting should be your last option as a choice in the kitchen. The reason is that it is virtually impossible to keep it clean and/or sanitary. If you choose to have carpeting in your kitchen, use a commercial grade only, preferably with an antibacterial agent treatment. The antibacterial treatment will help to prevent undesirables

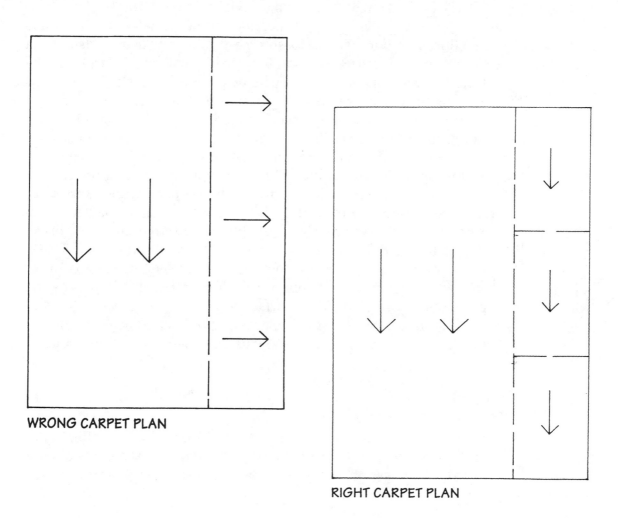

WRONG CARPET PLAN

RIGHT CARPET PLAN

from growing in your kitchen. The low nap and durability of the commercial quality will at least allow for reasonable maintenance and performance. Both commercial and residential carpets range in price from $6 to $50 a yard. The best prices are usually for those carpets that a retailer stocks because they have purchased a large quantity for a reduced price. Be sure to ask if installation is included, what kind of padding will be used, and how much they will charge to take up your old carpet and dispose of it. You should expect to pay about $1 a yard for taking up the old and disposing of it. Also be aware that many retailers will charge for moving your furniture out of the way. It is better to ask in advance and take the opportunity to move it yourself if you don't want to pay extra.

Concrete is actually becoming a viable choice for kitchen (and more) floors, particularly in warmer and humid climates. The reason is that it doesn't mildew and it doesn't matter if it gets damp. Also, it is a great option if you struggle with allergies because none of those pesky critters (such as dust mites) enjoy living on concrete. It can be painted to look like anything you desire from tiles to whimsical patterns.

One technique recently developed is *staining*. There are two ways to accomplish this. If you know in advance that you want to use concrete as a finish material, you can have the color mixed in at the time of installation. However, you can actually apply a staining product afterward. It is a product that penetrates the porous exterior surface to give a deep wonderful color. It is available in stonelike shades of tan, green, black, red, and blue. It is suitable for concrete, unglazed brick, and tile. It will not fade, chip, or peel. Kemiko Products (903-587-3708) is one manufacturer I recommend. They also have a wax available that will give the concrete a beautiful patina that enhances the depth of color while providing surface protection from stains. In addition, they have a penetrating sealer that provides easy maintenance. Some very expensive homes are now using concrete as a floor choice.

Lighting

Lighting is a very important aspect of our lives. It affects our mood, vision, perceived beauty, and color rendition. The study of light is one of the most complex sciences and is further complicated by the fact that light can be a combination of both artificial and natural. Obviously the amount of natural light in a home is controlled by the number of windows, skylights, and other natural light sources, and their relationship to the direction of sun exposure. Which unless we start adding windows and skylights, we really cannot effectively change. However, we can do a lot to increase and control artificial light.

As a designer, I always start with the basic floor plan. I feel it is essential to know what and where virtually everything will be placed into a space before I can begin planning lighting, electrical needs, or even finish products. Even if you don't yet own furniture for a particular space, plan ahead. I even take into consideration placement of artwork and other accessories. That way I avoid having light switches or other electrical outlets end up in the way. In my own home I lowered the placement height of several light switches specifically to accommodate the placement of large artwork.

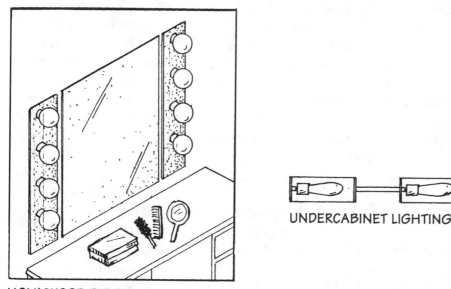

HOLLYWOOD BULBS

UNDERCABINET LIGHTING

FLUORESCENT LIGHT FIXTURE FOR UNDER CABINET

There are basically three types of lighting requirements: *ambient, accent,* and *task.* Ambient lighting is the most diffuse. Its job is to fill the entire space with soft illumination. Accent lighting is just that—focused to highlight special areas or objects. Task lighting is specifically suited for an individual task. The goal of task lighting is to provide the best possible light without a shadow or glare. It is important to address each of these areas when considering light for a space.

Light fixtures are specifically designed to meet these three needs. Ambient fixtures usually refer to ceiling-mounted styles that provide general room light. Some of the more popular choices are recessed cans or fluorescent fixtures that direct light upward. I suggest using a dimmer for your ambient light. That way you can change the "mood" with the turn of a knob. Ambient light can also be an integral part of your task lighting. A recessed light placed directly above a workstation can be effective at 100 percent capacity, while acting as ambient light at 50 percent capacity. Accent fixtures are usually such things as chandeliers, pendants, track lighting, or recessed lights. Their goal is to add extra light or focus to specific areas for such things as highlighting artwork or providing more light for dining. Task fixtures can be similar in styles or more specific. Recessed downlights and pendants installed directly over workstations can be quite effective as task lights. Examples of more specific task fixtures are desk lamps, music lamps, or even small reading lamps. Think in terms of the task at hand—to avoid casting your own shadow, install task lights slightly in front and above the surface you are working on.

Choosing the right type of lamp (bulb) is also critical to properly lighting your home. There are basically three types of lamps: *incandescent* (the one you're most familiar with), *fluorescent* (most likely what is in your office), and *halogen* (a more recent newcomer). *Incandescent* generates a warm yellow/red light that we find soothing and flattering. *Fluorescent* lamps' bad reputation of being a harsh light is no longer applicable. In fact, fluorescent lamps have a wide range of colors (shades) designed specifically to accommodate many uses. Bluish white is the one we are all most familiar with, but in fact it is available in warmer yellows, pinks, and even full spectrum. I have used full spectrum for many years in my design studio. It is the closest lamp to true sunlight. It is color correct (it does not change the true color of objects) and extremely efficient, using as much as one-third less energy than incandescent lamps. In addition, its lifetime is up to twenty times longer. *Halogen* lamps provide a very white light. They are a small lamp with tremendous power giving a clear white light. When first introduced, they were a major sensation because their small size gave them such flexibility. But as time has progressed we are discovering some serious drawbacks. They are expensive and can be damaged (burned out) by your touch. The most alarming problem is that they burn very HOT! They have been the cause of some serious fires. Do not use them near anything combustible. Please do not use them near children. The newer fixtures designed for use with halogen lamps must now provide a protective screen. However, this does not reduce the amount of heat that they generate. If you own an older halogen light fixture that does not have a screen, consider eliminating it, or check with the manufacturer to see if they can provide a screen. I do still believe halogen lamps, when used properly, can be very effective. They allow for some very creative uses. I recommend using a combination of lamps for the best balance of light.

Kitchens are probably one of the most complicated rooms to light properly. Which makes sense, since they are such complex centers of activity. Square ceiling-mounted fluorescent fixtures are the number one choice for ambient light in the kitchen. They are energy efficient, cool burning, and don't cast shadows. The second most popular fixtures for ambient light in the kitchen are recessed and track lighting, although both are more limiting than fluorescent. Light fans out from the light source in a cone shape. The amount of light diminishes as the beam spreads out. You can actually mathematically plot where and how much light any given source will provide to a surface. Obviously, the larger the lamp, the greater the beam spread. Incandescent also burns hotter than fluorescent. So, if you need a lot of them to light your kitchen, it can get pretty hot, especially if the oven is on.

Hanging lights are an excellent source of task lighting in the kitchen. By positioning them over an island, peninsula, or dining area, you can specifically light the space you need. Undercabinet lighting can provide additional task lighting and is available in both fluorescent and halogen styles. If choosing halogen, be sure to specify frosted lenses to diminish glare. Undercabinet lights should be placed at the lead edge of your upper cabinet. Otherwise, all the light will be directed to the backsplash (the vertical wall between your countertop and your wall-hung cabinets) and not the working counter area. If you choose to use undercabinet lighting, you can easily add plug molding (strips with

KITCHEN LIGHTING PLAN

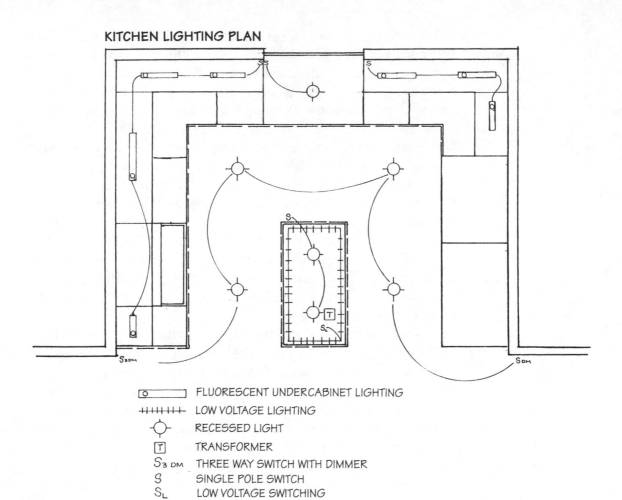

▭	FLUORESCENT UNDERCABINET LIGHTING
╫╫╫╫╫	LOW VOLTAGE LIGHTING
◇	RECESSED LIGHT
T	TRANSFORMER
$S_{3\,DM}$	THREE WAY SWITCH WITH DIMMER
S	SINGLE POLE SWITCH
S_L	LOW VOLTAGE SWITCHING

electrical outlets) every few inches. This can eliminate the need to have your backsplash area cluttered with outlets.

Bathrooms are the next most complex area to light—although, most people don't seem to give it much thought. Since this is where most of us begin our day, it would be nice to start with a flattering impression. To accomplish this you will need to carefully position lights around the mirror. Hollywood still has the best idea for this—"Hollywood-style" bulbs around the mirror are very effective. The next best option is to use strips of diffused light on either side of the mirror. The third most effective method of lighting at the mirror is to use fluorescent lamps with a diffuser overhead. Avoid recessed downlights

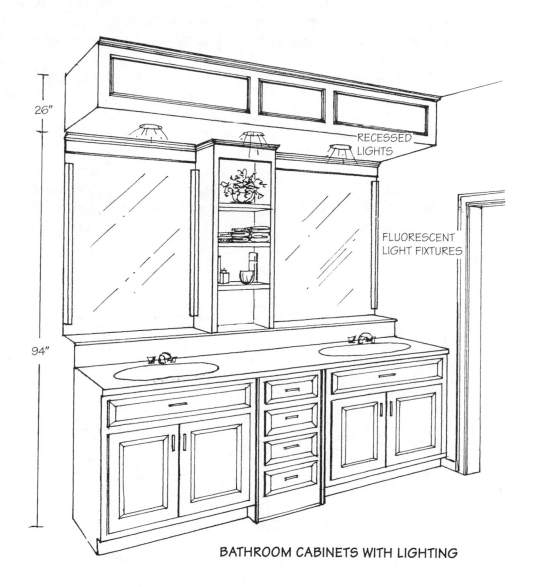

26"

94"

RECESSED
LIGHTS

FLUORESCENT
LIGHT FIXTURES

BATHROOM CABINETS WITH LIGHTING

overhead. They cast shadows that can make you look like you need another night's sleep. Warm-toned incandescents or fluorescent bulbs create the most flattering skin tone. Halogens provide clean, white light that mimics daylight. I also suggest adding a lighted magnifying cosmetic mirror at your vanity. This will allow you to choose specific types of light for specific situations such as day or evening makeup applications.

The balance of your bathroom should be lit with diffused lighting in strategic locations. My husband, for example, enjoys an evening shower in a dimly lit bath. By using a dimmer on the fixture above the shower, he can set the light to the mood he desires. I did the same with the fixture above the tub.

I usually design a lighting plan to provide good ambient light for the balance of the house. I am a firm believer in using dimmers to allow for choosing how bright or dim a room is, depending on the activity. I attempt to provide all ambient light with ceiling fixtures. In my own

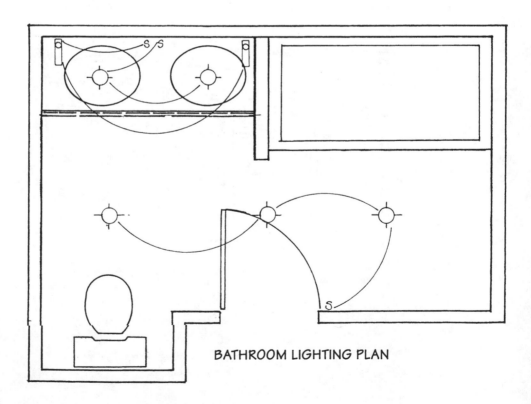

BATHROOM LIGHTING PLAN

Dressing areas are one of those places that need to have special consideration. Color-correct lamps are important if you want to be well dressed. How many people have had the experience of later finding that they've put on one black sock and one brown sock?

home, since I have vaulted (two-story-high) ceilings, I used recessed lighting for the majority of ambient light.

I also think that foyers or entryways should be lit to gently bring you into the house. This becomes a particular consideration in the evening. I think that there should be a gradual increase of light as you progress into the house. As you enter from the dark outside, the first light of the foyer should be subtle to allow the eyes an opportunity to adjust. Then as you progress farther into the house, the light can increase in volume. As a result, I used wall-mounted sconces that gently provide a glow of light on the walls of my foyer. Then an overhead fixture (on a dimmer) at the far end takes you up a notch in light volume. Finally, as you enter the great room, the lighting is at a level of volume for full vision. By using light creatively, you can completely change the mood of the room and your emotions.

It is also necessary to consider other activities that must be accomplished when planning lighting. For example, I always make sure that each space has additional lighting for such things as cleaning. I can't stand trying to clean, vacuum, or even dress in dingy or dimly lit spaces. So, in addition to ambient mood light, I make sure there is enough power for the basic mundane tasks as well.

I still like the idea of having a basic ceiling-mounted fixture in every bedroom. For a while, many builders had stopped providing them. A simple, inexpensive fixture can make lighting a bedroom so easy.

The most important aspect to lighting is planning for it. In terms of budget, I usually find that to do it right will usually take twice as much money as the builder has budgeted for wiring. That's because most builders will use lower figures for most aspects of a budget to bring the overall cost within a budget they think you will accept. The problem arises when you realize you cannot get what you feel you need or want within that lower budget. By this time, you've most likely already signed the contract to proceed with the project. However, I don't think you have to spend a fortune on lighting fixtures. Choose interesting decorative fixtures for where they will have the most impact—

The more specific you can be with a builder in terms of what you want, the closer to actual budget the estimates will be.

above a dining table, in a foyer, or even in a guest powder room. Everywhere else, be practical—find something pretty and inexpensive that provides the best kind of light for the task. Lighting fixtures can be found almost anywhere today. If you are working with a builder or electrician, ask if you can purchase your fixtures where he has an account. In many cases, the contractor will extend his discount to you, which can save a lot of money. Expect to pay about $15 to $25 for simple incandescent recessed fixtures. Low-voltage halogens can cost as much as $200 each. Fluorescent undercabinet lights start at about $18 per foot. Halogen undercabinet lights are $20 to $30 plus the cost of a transformer—$35.

Words of Wisdom

❀ If you can't afford to replace your old vinyl floor—paint it! Many paint manufacturers have developed primer (base coat paint) that creates a better bond with vinyl flooring. You can choose the technique and style of your choice. One company makes a kit to create the look of a brick floor with stencils and paint. It's called the Floor & Patio Paint & Stencil Kit ($50) by Plaid Enterprises, P. O. Box 7600, Norcross, GA 30091-7600. For information on other techniques check with your local paint store.

❀ Tile is heavier than resilient flooring—be sure your floor can handle the weight.

❀ One way to create the feeling of more room in a small home or area is to use the same flooring material throughout. This continuous element gives the illusion of continuing space. It's the same theory I use when dressing—by wearing the same color from neck to toe I create the illusion of being *much* taller than my actual height of five foot two.

❀ Benjamin Moore paints offers Web site how-to information for a variety of projects such as painting your vinyl floor to look like faux terra-cotta tile. Web site: www.benjaminmoore.com.

❀ It is not the responsibility of the carpet installers to plane down your doors. If they don't fit after the new carpet has been installed, this is your problem to fix.

❀ If you choose to use a lighting designer, you can call for recommendations to the American Lighting Association (800-274-4484).

❀ Fluorescent lamps last ten to twelve times longer than incandescent.

❀ Specify nonslip flooring such as matte-glaze or textured when using tile.

❀ Keep a fire extinguisher in your kitchen (but not near the stove). Instruct everyone in the house how to use it, and have it recharged as needed.

❀ For safety's sake, be sure to install fixtures that are rated for wet locations in enclosed shower or shower/tub areas.

❀ Install a smoke detector *near* the kitchen but not *in* the kitchen, where ordinary smoke and heat can set it off. Some are now equipped with a nuisance silence button—it allows you to deaden the sound at the touch of a button.

❀ Plan in advance for a safe and secure location for sharp knives, especially if you have small children.

PART TWO

Interior
Renovations
and Products

4

KITCHENS: PRICING, PLANNING, CABINETS, AND COUNTERTOPS

The kitchen has always been the "heart of the home." Today's modern kitchen is no exception. I find it interesting that *today's* kitchen is looking more and more like *yesterday's* kitchen. Think about it—remember the old colonial kitchen, with the warmth of the center fireplace—it was more of what they called a "keeping room." Everything happened here: cooking, reading, warming, living. In fact, in the old log homes of yesteryear, the kitchen constituted most of the home. If you've been to an open house or a Parade of Homes recently, you will have seen the latest trend—"keeping room" kitchens! A fireplace with a cozy loveseat and chair, a sunny corner for dining, a counter seating area for chatting with the cook, a computer station, a central security video/alarm/watch guard center to see who is coming and going, a wine/beverage center, and even a stove! The fact is, we are doing more living in our kitchens than anywhere else. It's no wonder we put so much money and emphasis on this one room!

Materials and Budgets

Money is always the first subject of discussion when it comes to redoing a kitchen. The first question is *how* should I spend the money I have budgeted? The second, what will I get for my money? The "rule of thumb" says never spend more than 10 percent of the total value of your home on the kitchen. In other words, if your home is valued at

69

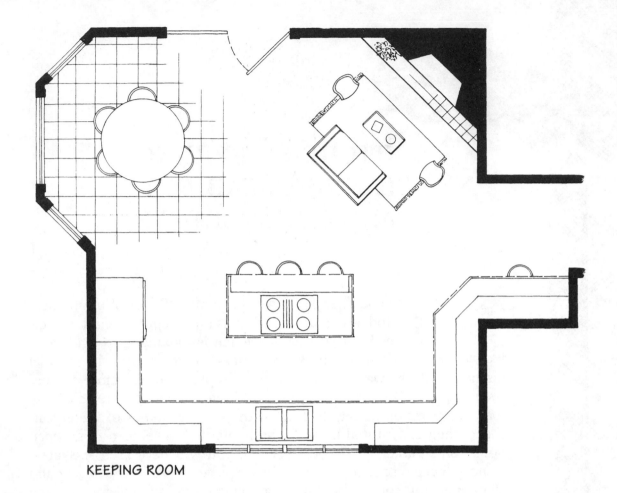

KEEPING ROOM

$100,000—spend $10,000 on the kitchen. Why? Because when it comes to selling your home, realistically this is all it will be valued at. But many people are spending more. I have seen a lot of kitchens in recent history that went way beyond this 10 percent rule. I think we are putting more emphasis on the value of the kitchen in general. Certainly, resale value will be directly related to the desire of those buying the home. A gourmet cook who entertains in the kitchen will be willing to pay more for a dream kitchen than the one who always makes reservations for dinner.

Also, this 10 percent rule does not necessarily apply to redos. When redoing a kitchen, you must take into consideration the "undoing" as well! It costs money to rip out an old kitchen and cart away the trash. How much you ultimately decide to spend will depend on how

A CONTEMPORARY-STYLE KITCHEN

long you plan on staying in your home, what your expectations are, and of course, how much money you can afford to budget. Below is a list of average statistics that will give you some basis for determining your own budget.

According to a survey by the National Kitchen and Bath Association (NKBA) the average cost of a kitchen remodeling project including the cost of a certified kitchen designer in 1996 was $22,100. Of that total budget, the average consumer spent:

- ❀ 48 percent on cabinets
- ❀ 16 percent on installation and labor
- ❀ 13 percent on countertops

Kitchens: Pricing, Planning, Cabinets, and Countertops • **71**

- 8 percent on appliances
- 6 percent on design
- 4 percent on fixtures and fittings (sinks, faucets, etc.)
- 4 percent on flooring
- 1 percent on miscellaneous extras

I find it interesting that 60 percent of all kitchen remodeling is done by households with heads aged forty and older. Overall a kitchen redo can cost anywhere from $16,000 for basic to $50,000 for top of the line. Here are some guidelines as to what you can expect to get for your money.

$3,000 TO $5,000

- Reface existing cabinets (Cover the outside face of your cabinets with laminate and replace doors.)
- Replace countertops with *postform* laminate (Postform tops are pre-made with a curved edge and are available by the foot.)
- New builder's base grade vinyl floor

$5,000 TO $10,000

- Same as above with the addition of new appliances and a custom laminate countertop

$10,000 TO $20,000

- No change in floor plan or layout of kitchen. You will be able to replace approximately fifteen to eighteen cabinets with builder quality stock cabinets. Stock cabinets are manufactured in large quantities in basic standard sizes and a few basic finishes such as: laminate (i.e., Formica brand), thermafoil (a scratch-resistant thin laminate that is "shrink-wrapped" to a wood core, creating the look of white painted cabinets), oak, or pine.
- New laminate countertops
- Minimal changes in plumbing
- Basic electric upgrade
- New basic (freestanding) appliances

- ❀ New resilient flooring
- ❀ Ceramic tile backsplash

$20,000 TO $30,000

- ❀ Make some changes to floor plan
- ❀ Upgrade to medium-grade materials and semi–custom cabinets. Semi–custom cabinets are a combination of stock and custom with added options such as crown molding, trim, and interior fittings and accessories
- ❀ Replace flooring with hardwood, ceramic tile, or laminate flooring
- ❀ Upgrade countertops to solid surface materials, tile, marble, or granite

$30,000 TO THE SKY IS THE LIMIT . . .

- ❀ Enlarge kitchen, take out walls, add a room, deck, or porch
- ❀ Custom cabinetry made specifically for your kitchen and your needs, with expensive door styles and any option on wood you prefer
- ❀ Professional-style built-in appliances including the latest greatest trend
- ❀ Skylights, French doors, new windows and doors
- ❀ Additional freezer, refrigerator, or second cooking/baking station

So, how do you determine where you fit in? In my experience, most people have some idea of what they want their kitchen to look like. They know, for example, if they want the rich warmth of deep cherry cabinets, or the lightness that white-painted cabinetry would bring. They usually know the basic colors they want to work with. They may even have some ideas about basic needs, such as adding a microwave or an island with a sink. But for most, the details of how we get to the final picture are a bit foggy.

What Am I Looking For?—Making a Wish List

To lift some of the fog, it is important to begin with a list of expectations and goals. Here are some questions to begin your list.

LIST A—THE BASICS

❀ For what kind of items do you need additional storage? Dishes, glasses, pots and pans, cereal boxes, canned goods, baking cookware, pet food, trash, recycling products, toaster oven, or the coffeepot? It's important to be specific in order to adequately design the space to fit your needs. One of the easiest ways to add additional storage is by using taller wall-mounted cabinets. This will add at least one shelf to the inside of each cabinet, and it's a great place to put those seldom-used items.

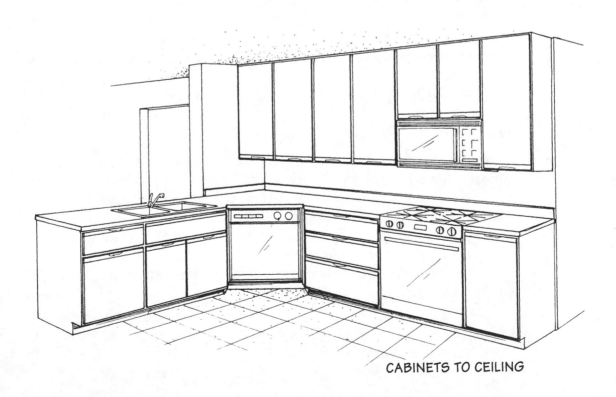

CABINETS TO CEILING

❀ Which of your appliances needs to be replaced? Which can be eliminated? What additional appliances do you need? (A freezer, microwave, additional oven?)

❀ Do you need more counter space? Consider space-saver appliances that can be mounted underneath the upper cabinets to conserve precious counter space.

❀ Evaluate your lighting needs. Is your kitchen dark, even during the day? If so, you may want to consider additional windows and/or skylights.

❀ How much cooking do you really do? Is there more than one cook in the house?

❀ Do you need a place for a computer? Where do you want the telephone message center and calendar located?

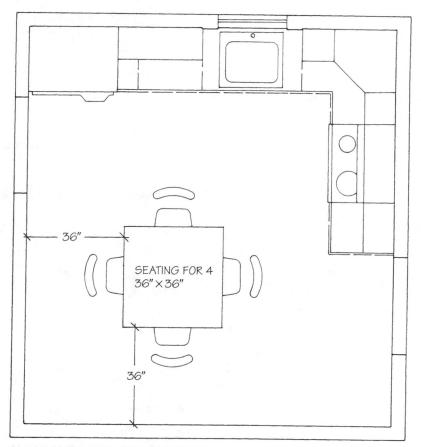

CLEARANCE FOR FOUR DINERS

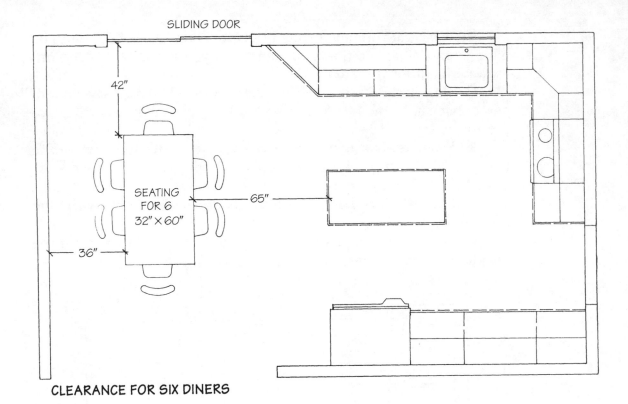

SLIDING DOOR

42"

SEATING
FOR 6
32" × 60"

65"

36"

CLEARANCE FOR SIX DINERS

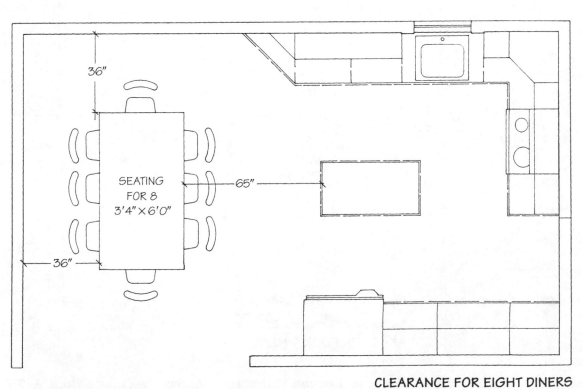

36"

SEATING
FOR 8
3'4" × 6'0"

65"

36"

CLEARANCE FOR EIGHT DINERS

❀ Is there sufficient space to accommodate a table and chairs for eating in your kitchen? My mom's kitchen is small, and when my granddad moved in, it was very difficult to accommodate his wheelchair at the kitchen table. Remember to consider long-term needs when planning a redo.

❀ Where do you want the laundry? Often it is located near the kitchen for convenience, but the problem is that it's noisy, which means you may want to consider other options.

❀ How tall are you? Yes, this is important because more and more, kitchens are being customized to meet our physical needs. For example, if you are petite (under five two) you may want to lower your cooking work area from a standard thirty-six-inch height to thirty inches.

❀ What are the things you like about your kitchen?

❀ What is it that bothers you the most about your kitchen?

❀ If you could change the traffic flow, how would you make it different?

❀ Consider a place for your furry friends—plan a pet-food station.

❀ Do you need more than one sink location?

❀ Do you need a place to keep your cookbooks? Now is the time to plan for a bookshelf and/or a cabinet space for all those tasty recipes.

Be honest with yourself. If you are not a neat-nick, then be sure the majority of your storage spaces are hidden behind doors. Sliding wire basket systems or pull-out shelf/drawers are ideal for being able to quickly see what's where. A pantry closet can double as a cookbook keeper for all those cookbooks that are not so pretty to look at.

LIST B—HOW FAR SHOULD WE GO?

This list deals more with the scope and extent of the project.

❀ Are you happy with the overall floor plan? If so, you may want to consider *refacing cabinets and new countertops* rather than new cabinetry. Obviously, this will to some extent depend on the condition of the interior of the cabinets. (See Refacing, later in chapter.)

❀ Is it necessary to enlarge the entire kitchen, or is the size sufficient?

BLENDING OLD AND NEW CABINETRY

❀ Are new windows and doors necessary? Do you want to add a door or doorway? Many of my clients have added *pocket* doors, which give them the option of closing the kitchen off without taking up precious wall space.

❀ Do you need a new floor? Depending on your new choice, this will probably mean a new subfloor. If you choose tile, additional support may also be required.

❀ Do you want to open up the kitchen to the adjoining room? This too is popular, especially if you have a family room near.

The next step I recommend taking is to do some serious "window-shopping." So much can be learned from this experience. Don't be

BEFORE RAISING ROOF OF KITCHEN AND RENOVATING

AFTER RAISING ROOF AND RENOVATING

afraid to shop out of your budget. This is your opportunity to explore all the options, even if you can't afford them. You will be amazed at the kind of valuable ideas you can find. You may for example decide to splurge on a built-in food processor. For me, this was money well spent. It cost only a few hundred dollars. Having it built into my countertop made it convenient and saved valuable counter space. Take notes and collect information for later evaluation.

While window-shopping, use this opportunity to speak with and interview possible designers and/or contractors. Be honest with them. Let them know you are just beginning to consider a renovation and are exploring your options. Model homes, home centers, builder shows, and kitchen cabinet manufacturers are ideal places to look.

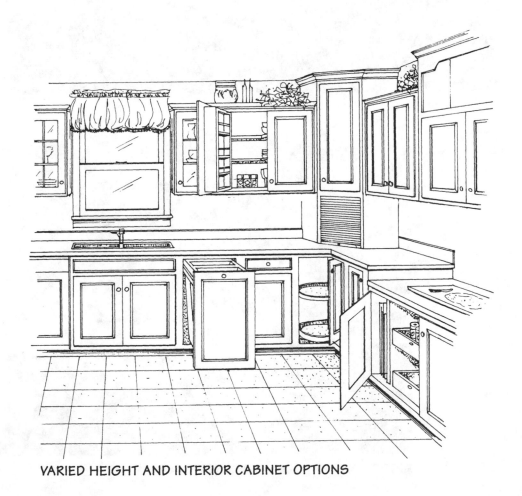

VARIED HEIGHT AND INTERIOR CABINET OPTIONS

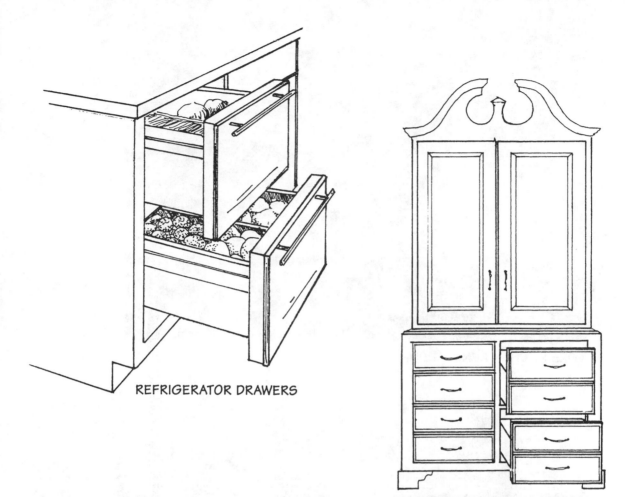

REFRIGERATOR DRAWERS

REFRIGERATOR IN AN ARMOIRE

Be sure to include appliance dealers as well. This is one of the most important aspects of your kitchen. Refrigerators are a good example of how trends are changing all the time. In the late 1950s and early 1960s a few manufacturers introduced in-cabinet and drawer refrigeration units. Well, they're back. But this time, they have individual temperature and moisture controls for specific tasks such as storing wine, freezing ice cream or keeping vegetables. They are even now designing refrigerators to look like furniture pieces such as an armoire or eighteenth-century highboy cabinet.

The next place to focus is on evaluating specific elements of the kitchen, since it is virtually impossible to begin budgeting without it.

BEFORE KITCHEN RENOVATION

AFTER KITCHEN RENOVATION

The Plan

Overall floor plans are changing in the kitchen. Until recently, the "work triangle" was the most important aspect of the layout. In the work triangle, the sink, stove, and refrigerator formed a triangle. Today, there is less importance placed on the triangle and more focus on interrelated working stations, each with a primary function. A workstation can be created wherever there is enough counterspace next to a major appliance. With the popularity of islands, oftentimes several smaller workstations can be appropriated here. This allows for more freedom and flexibility in design and an opportunity to create a space that functions specifically to your style.

One of the most popular workstations is the cooking center. A cooking center can become the focal point architecturally in the kitchen. Using custom-hooded ranges with elaborate trim and molding, much like a fireplace mantel, they can be exquisite. Another often used workstation is the baking center—often using marble as a countertop for rolling out dough. One word of caution—don't let someone

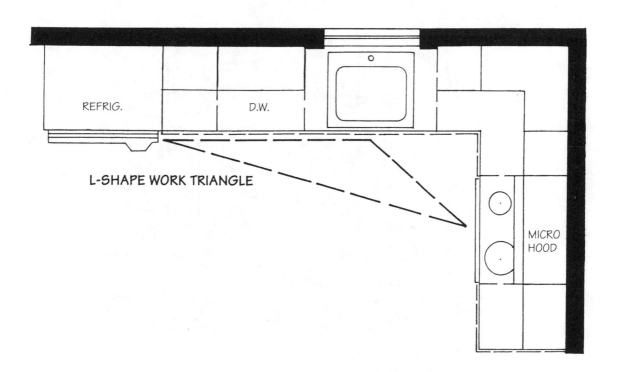

L-SHAPE WORK TRIANGLE

REFRIG.

D.W.

MICRO HOOD

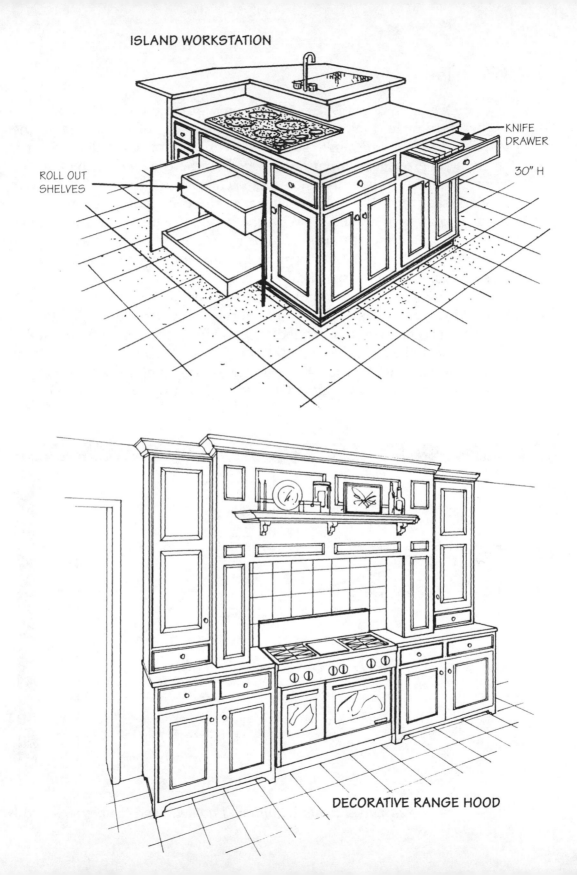

ISLAND WORKSTATION

KNIFE
DRAWER

ROLL OUT
SHELVES

30" H

DECORATIVE RANGE HOOD

DECORATIVE RANGE HOOD

talk you into more than you need. Be honest with yourself. Are you a gourmet cook, or is throwing a "prepared" meal into the microwave the extent of your culinary expertise? Remember, this is *your* house, not Julia Child's.

If you have an old *galley*-style kitchen, and the basic shape cannot be changed, then consider segmenting it. By dividing the long space into perhaps three separate areas, dedicated to specific functions, you can create the illusion of a wider, better-proportioned space. You could set the working kitchen in the center, while designing a butler's pantry at one end and a mudroom/storage area at the other. The use of transoms above doorways can add visual definition without completely closing off the spaces. Use one continuous floor treatment throughout. Choose the same wall treatment for the two end spaces with a complementary one for the main kitchen work area. You could also create some diversity of interest in your cabinets by using the same style throughout while using a combination of finishes.

As the kitchen has evolved, some of the old standard measurements no longer are etched in stone. Included below are some suggested measurements from the National Kitchen and Bath Association to be considered in your planning.

DECORATIVE KITCHEN CABINET TREATMENT

DECORATIVE RANGE HOOD IN CORNER

OVER, RANGE BUILT INTO CABINETS

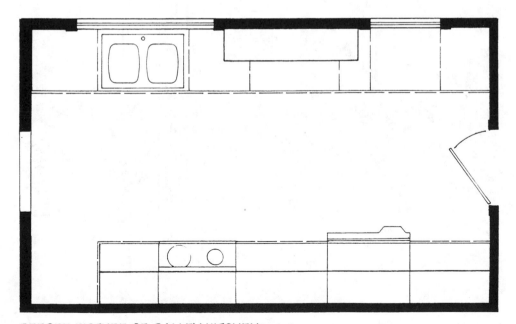

BEFORE PICTURE OF GALLEY KITCHEN

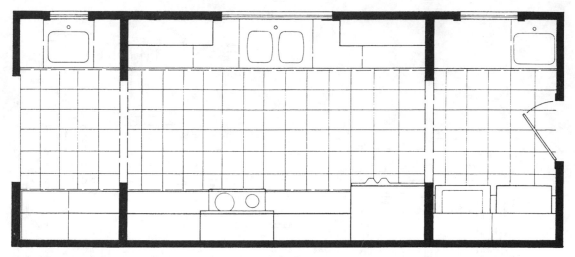

AFTER PICTURE OF GALLEY KITCHEN

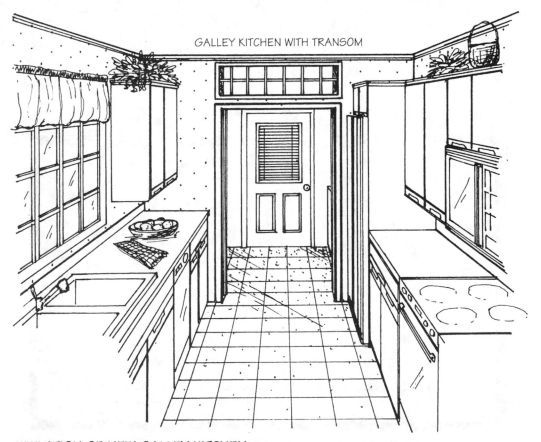

GALLEY KITCHEN WITH TRANSOM

ELEVATION OF NEW GALLEY KITCHEN

NKBA GUIDELINES FOR COUNTERTOP HEIGHTS AND CLEARANCES

Work Area

- ❀ Requires at least 36 inches of continuous countertop.
- ❀ At least one such area should be next to the sink.
- ❀ If you have more than one cook, each needs at least 36 inches of continuous countertop.

Cook Top

- ❀ In an open-ended countertop run, you need 9 inches of countertop, 15 inches on the other side.
- ❀ In a close-ended run (between walls), you should have at least 3 inches between the cooktop and a flame-retardant wall. And at least 15 inches on the opposite side.

Bake Center

- ❀ The optimal counter space for baking is a minimum of 36 to 42 inches wide.
- ❀ The counter surface should be 4 to 6 inches lower than the height of your elbow when bent at a ninety-degree angle.

Oven

- ❀ If the oven is located near a main traffic area, you should have 15 inches of counter space above or on one side of it.
- ❀ If not, you can have 15 inches of space next to or across from the oven no more than 48 inches away (for example, an island).

Refrigerator

- ❀ You need at least 15 inches of counter space on the nonhinged side of the refrigerator.
- ❀ A side-by-side refrigerator should have 15 inches of counter space on either side.

❀ Or there should be fifteen inches of counter space across from the refrigerator no more than 48 inches away.

Countertops

Standard countertops are 36 inches high, but as I mentioned earlier, this is changing. Often this height is adjusted to meet your height. Most people are comfortable with a counter height of 2 or 3 inches less than the distance from floor to elbows when standing. It is also recommended to have different heights for different activities. The NKBA suggests putting baking centers at 30 inches, desks at 29 inches, and breakfast bars at 42 inches.

NKBA GUIDELINES FOR COOKING CENTERS

❀ Cooking surfaces (stove tops) should not lie below an operable window unless the window is at least 3 inches behind and 24 inches above the appliance.

❀ Microwave ovens should be placed so the bottom of the oven is 24 inches to 48 inches above the floor.

❀ Work counters should be at least 16 inches deep, and wall hung cabinets should be at least 15 inches above the work surface.

Nitty-gritty—Cabinets and Countertops

Now it's time to start looking at all the different elements that ultimately make up your kitchen. I'm going to attempt to cover everything in a reasonable way, giving you basic information while also throwing in some of my personal thoughts on the subjects.

CABINETS

Let's start with the most expensive part of your kitchen, the cabinets. This is also the area where you have the greatest diversity and choice of price and quality. I have done complete custom-designed and handmade kitchens in the $40,000 to $50,000 range. And I have designed beautiful and creative kitchens from stock cabinets for less than

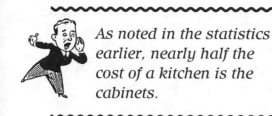

As noted in the statistics earlier, nearly half the cost of a kitchen is the cabinets.

$10,000. My own kitchen is a stock cabinet to which I added stock moldings to the top and bottom, and a little "gingerbread" to create what appears to be a custom kitchen. The salesman I worked with had never tried such a feat before. He was not too thrilled when we began. But as we progressed, he told me how much he had learned and was thrilled with the outcome. The key to using stock materials and "customizing" them is a good installer. It takes a *finish* carpenter to make all the pieces fit as if they were made for one another.

There are basically two types of kitchen cabinets: *base* and *wall*. Base cabinets sit on the floor and are generally 34½ inches high. (Once a countertop has been added, they will be 36 inches high.) Wall cabinets are attached to the wall. They vary in height. In addition, there are

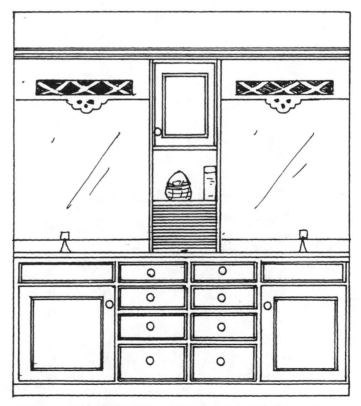

STOCK CABINETS WITH TRIM TO DRESS THEM UP

Kitchens: Pricing, Planning, Cabinets, and Countertops • 91

basically two types of construction for cabinetry: *face frame* and *frameless*. Face frame is the more traditional style. The frame is visible even when the doors are shut. The hinges may or may not be visible. Frameless construction means the doors completely hide the frame of the cabinet. Hinges are also almost always hidden. This is generally used for more contemporary styling.

    Stock cabinets are generally manufactured in widths beginning at 9 inches and increasing in size in 3-inch increments. The most common width is 30 inches. A 30-inch stock-base cabinet will range in price from $135 for a flat-panel laminate to $250 for recessed panel in oak. For more intricate styling in the next grade level of wood, such as cherry, the price is about $350. Stock wall cabinets, 30 by 30 by 12 inches, run about $100 to $220. For cherry, the price is about $275. Stock cabinets will almost always have a *center stile,* which is a vertical support in the center, between two doors. The disadvantage to this is

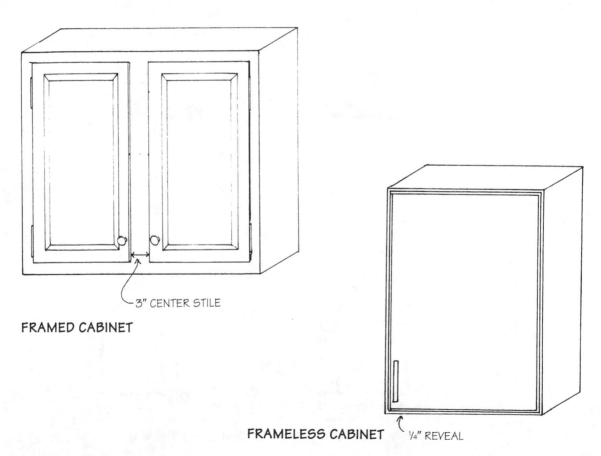

3" CENTER STILE

FRAMED CABINET

FRAMELESS CABINET ¼" REVEAL

that it makes it more difficult to get larger items in and out of it. But the cost is less than you would pay for semicustom. You will need to weigh the advantages versus disadvantages yourself.

Usually stock cabinets are "in stock," which means they are available for immediate delivery. However, it is not uncommon to "special order" stock cabinets specifically for one job. This will usually take ten to twenty days.

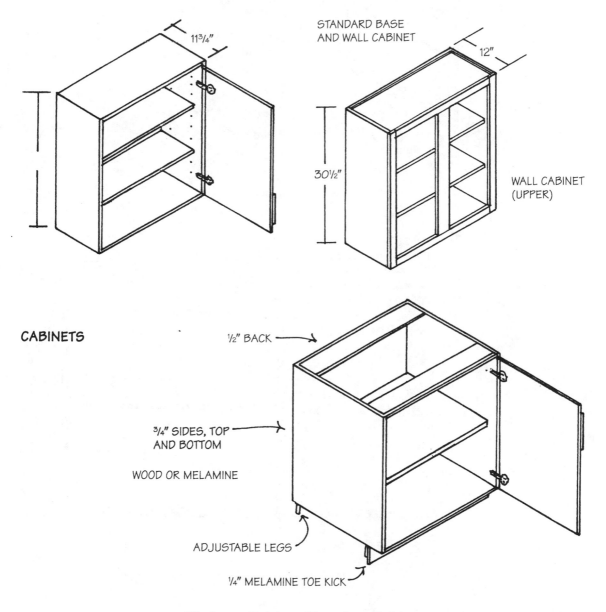

STANDARD BASE
AND WALL CABINET

11¾"

12"

30½"

WALL CABINET
(UPPER)

CABINETS

½" BACK

¾" SIDES, TOP
AND BOTTOM

WOOD OR MELAMINE

ADJUSTABLE LEGS

¼" MELAMINE TOE KICK

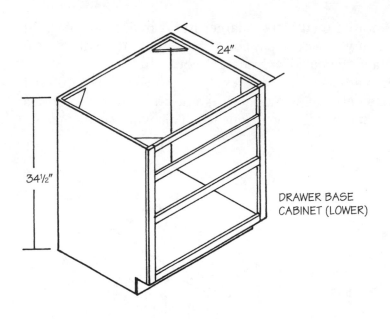

34½"

24"

DRAWER BASE
CABINET (LOWER)

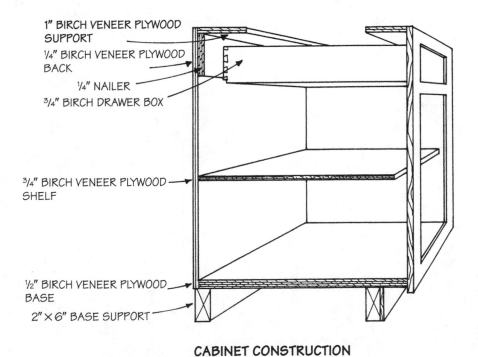

1" BIRCH VENEER PLYWOOD
SUPPORT

¼" BIRCH VENEER PLYWOOD
BACK

¼" NAILER

¾" BIRCH DRAWER BOX

¾" BIRCH VENEER PLYWOOD
SHELF

½" BIRCH VENEER PLYWOOD
BASE

2" × 6" BASE SUPPORT

CABINET CONSTRUCTION

Semicustom cabinets are generally a higher quality. You will find ball-bearing, full-extension drawer glides as a standard feature. In addition, you will have more options, which include more variety in door styles and the size, shape, and function of cabinets. Also, you will have interior accessories and options to choose from. They will be gap free with solidly fitting and sliding drawers. Trim and detail pieces will be made to fit. Delivery time is usually four to six weeks.

Custom cabinets are exactly what they sound like. Each cabinet is individually designed and made for the specific place in which it will be placed in your kitchen. They usually cost twice as much as stock cabinets. The standard features on custom cabinets include, but are

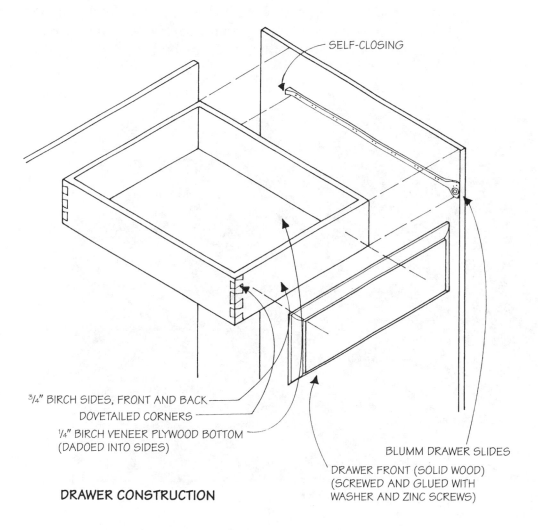

SELF-CLOSING

3/4" BIRCH SIDES, FRONT AND BACK
DOVETAILED CORNERS
1/4" BIRCH VENEER PLYWOOD BOTTOM
(DADOED INTO SIDES)

BLUMM DRAWER SLIDES

DRAWER FRONT (SOLID WOOD)
(SCREWED AND GLUED WITH
WASHER AND ZINC SCREWS)

DRAWER CONSTRUCTION

not limited to, unlimited choice of materials and finishes, dovetailed drawer joints, matching appliance panels, three-quarter-inch-thick shelves, and finely crafted details. A custom kitchen is more like a fine piece of furniture.

Cabinet doors are the real character and personality of a kitchen. Some have a rustic or country look. Others will be contemporary and sleek. The basic material choices for doors are wood veneer, solid wood, laminate, and paint. But that is just the beginning—glass doors, tin inset panels, raised panels, recessed panels (square or curved), and beaded panels are some of the other options you may find. Using a combination of door styles and finish colors can give you the ability to express your style in a unique way. Staggering the heights of cabinets also can add to a custom look without going completely custom. Open shelving, plate and wine racks, and a decorative range hood are also easy to incorporate without breaking the budget.

DOOR STYLE CREATES THE CHARACTER OF THE ROOM

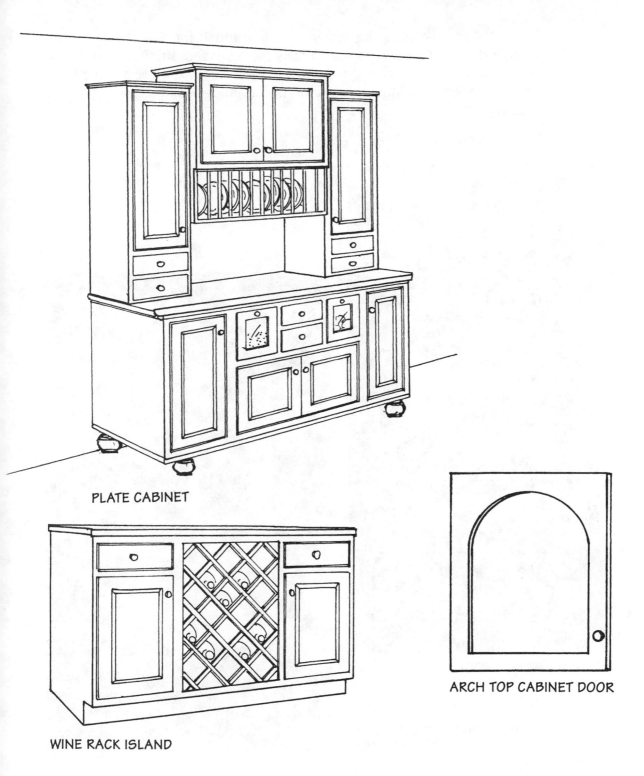

PLATE CABINET

WINE RACK ISLAND

ARCH TOP CABINET DOOR

When choosing the material and/or finish for your kitchen cabinets there are several things to consider—upkeep, lifetime, durability, and how it will react and respond to use over time. Here are the basic options, expectations, and maintenance guides for the most popular choices.

KITCHEN CABINET STYLES

KITCHEN CABINET STYLES USING
COMBINATION OF FINISHES

Wood Veneer

A thin layer of wood is glued to a substrate panel. This technique is often used for center panels of doors. It is durable but over time can warp and crack if near lots of heat and moisture.

Uses. It is ideal when you want the grain to be matched from one piece to the next. It is flexible enough to be curved. But it cannot be routed or carved.

Care. Use a mild diluted detergent and a damp cloth. Avoid abrasive cleansers.

Hardwood

Many woods are considered hardwoods. Maple, oak, and cherry all fit this description. All woods will react to temperature and moisture changes, which causes warping or shrinking. Shrinking can cause gapping. You may find gap in the winter, and then none in the summer. Also be aware that some woods will lighten with sun exposure, while others will darken with age. If a portion of your cabinets gets direct sunlight from a window, that portion will change color over time. You may also find that it dries out and needs to be refinished sooner.

Uses. Unlimited choices here—can be used in almost all situations.

Care. Mild diluted detergent and a damp cloth. Avoid abrasive cleansers. Occasionally use a no-wax polish.

Painted

This is usually done to hardwood, with a fine grain. The type of paint and style of paint will greatly determine the look. One of the newest popular "old" looks is a rustic, worn-edged style. With exposure to sun and temperature change, you may see cracking over time.

Uses. Again, the flexibility of paint makes it a good choice for almost any style. It can be glossy and sleek or it can be worn and pickled in appearance. Mixing a stained wood with a paint can be a beautiful way to create a custom-designed look.

Care. Avoid abrasive cleansers. Only use a diluted detergent with a damp cloth.

Softwood

Pine is the most common softwood in use for kitchen cabinets. In the past, it was usually dark and rustic. Today, it is becoming more stylish with a lighter honey-color stain and a finer finish, making it very appealing for today's home. Softwoods are more likely to scratch and dent. And like all woods, heat and moisture will affect it, causing warping and cracking.

Uses. About the only thing I do not recommend trying with softwoods is detailed carving. If carved, the wider grain of these woods makes them susceptible to cracking and even breaking at a grain point. It can be painted or pickled, creating a warm patina look. Do expect the knots to show through; this is part of the beauty of pine.

Care. Use mild diluted detergent with a damp cloth. Occasionally use a no-wax polish.

High-Pressure Laminate

This is made of layers of compressed resin and paper that are glued to a substrate. It's popular because it's easy to maintain. It's the same stuff your countertop is made of. The biggest problem with laminate is that once it's chipped, you cannot fix it without replacing it. Also, over time, the whites can yellow. However, this is true of almost all whites, whether laminate or painted.

Uses. This is best used for flat-panel doors. It cannot be routed or carved. It can be complemented with a contrasting color to create interesting appeal.

Care. Do not use abrasive cleansers. Be sure to keep edges dry, otherwise they will lift (unglue).

Low-Pressure Laminate

Often called melamine, it is thinner and less durable than high-pressure laminate. It is also less expensive.

Uses. Because it is less durable, it is mostly used for the box of a cabinet instead of doors and drawers. Again, it works as a flat-panel surface and cannot be routed or carved.

Care. Water or nonalcohol base cleanser. Keep the edges dry, otherwise it tends to lift (unglue).

Polyester

This is one of the best technological developments. It is considered one of the most durable finishes around. Basically, it is a polyester paint. Which makes it strong and stable. It is much less prone to scratching and denting. But once it is scratched or dented, it cannot be repaired.

Uses. Ideal for any style, with a full range of colors. It is available in both gloss and matte finishes. The whites seem to stay whiter longer.

Care. Use water or a diluted nonabrasive cleanser. Do not buff matte finishes.

Thermafoil

Another technological breakthrough, it is a heat-activated vinyl composition that is basically heat-shrunk to conform to any configuration. Thermafoil shrink-wraps around the entire door. The biggest advantage is a surface whose maintenance is similar to laminate but has the look of paint. Color is more colorfast. (My own kitchen cabinets are white Thermafoil—I love them.)

Uses. Because of its unique ability to be formed to any shape, it can be applied over carved and routed surfaces. It is available in various colors, wood grains, and textures.

Care. Use a nonabrasive cleanser or a damp cloth. Do not allow it to get hot—it will delaminate.

Painting Existing Cabinets

One of the most frequently asked questions is, Can I paint my existing stained wood cabinets? The answer is yes. Whether you decide to do this "in-home," or to have doors and drawers removed and painted "off-site" will depend on your long-term goals and needs. Today's technology makes painted stained wood cabinets far more durable than in the past. The use of sure-grip primers that give the paint staying power is the key. If I have the option, I prefer to have doors and drawers painted off-site, in the controlled environment of a painting/drying chamber. This is more expensive, but you can cut some of the labor costs by doing the removal yourself.

One painting technique that I have used successfully involved using a decorative painting technique on the upper cabinet door insets (the panel on the center of the doors). First I painted all the cabinets an off-white color. Then just on the insets, I used a peachy-pink combination to create the look of marble. As a complement, I used a similar color countertop. It was beautiful, effective, and reasonable in price. This kitchen went from looking like a dark hole to wonderful in just a few days.

OLD KITCHEN WITH NEWLY PAINTED UPPER CABINETS USING MARBLEIZED TECHNIQUE IN CENTER OF DOORS

COUNTERTOPS

The countertop is the real workhorse of the kitchen. It must take the most abuse while still maintaining its beauty. Fortunately, it's the one area where the choice of materials has grown the most. In the past you basically had a choice of laminate or tile. Today, the list is nearly endless. The use of a combination of different materials is a popular way to add your signature. Contrasting materials and/or colors for the edge treatment of a countertop is also a great way to create interest easily. I used stock molding to edge my laminate top. It gives the look of a solid surface material (like Corian) without the expense.

Just because something is popular or looks good, does not mean it will be appropriate for you.

By using the right material for the job, you can add beauty and years of life to your countertop and your kitchen. With all these new choices, the biggest problem I have seen is improper installation techniques and specification. One manufacturer, Corian, has dealt with this by requiring special training for each of its dealers. This has reduced the number of dealers but eliminated a lot of problems.

And all materials have their drawbacks. The best advice I can give you, is always to go see an installation of a previous client before choosing your installer/contractor. Talk with them and ask what, if any, problems they are having. Find out if they wish they had chosen another material. Too often, we let our emotions make the choice. In the kitchen, practicality should be the first reason for any decision.

Here are some of the more popular options for countertops:

Laminate

Basically there are two types—standard and color-through (solid color). Most laminate manufacturers have specialty protective finishes available. They were originally developed for use in hospitals and laboratories. I have successfully used them in the home in areas where the countertop is subjected to more than usual abuse.

Standard laminate is made of several layers of resin-impregnated paper fused under pressure and heat with glue to a substrate base. It has an endless range of designs, finishes, and patterns. It can be had in matte or shiny, smooth or textured, wood grain, metallic, and virtually any color you can imagine. When seamed, it will have a black edge (or line). It does scratch and was not designed to be used as a cutting surface. Nor was it designed to serve as a hot-plate rest—it will burn and melt. The bad news is, it cannot be repaired. If it chips, scratches, or burns, it will have to be replaced. The good news is that it is the most affordable choice. Glossy and/or textured laminate is not a good choice. The reason is that they both wear off. The gloss gets dull and the texture wears smooth.

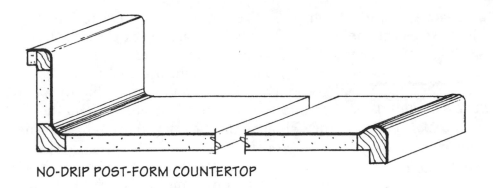

NO-DRIP POST-FORM COUNTERTOP

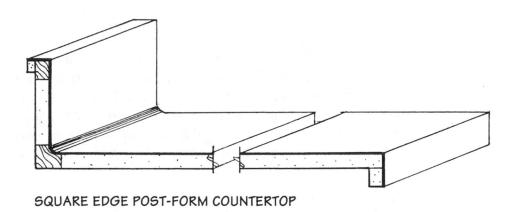

SQUARE EDGE POST-FORM COUNTERTOP

Color-through laminate is basically the same as standard laminate except the color runs all the way through the product. This means that seams and edges will not be black. Also, it is a little less obvious when scratched. It costs roughly twice as much as standard laminate. It too can have a decorative edge added. Most manufacturers have an array of stock, prefinished edge molding available.

Care. Use soap and water or any nonabrasive cleanser.

Solid Surfacing

Corian, Swanstone, and Avonite are the most popular brands. This plastic/resin is cast from a liquid into sheet form. As a result, it can be bonded to itself and formed into almost any shape. Seams are invisible when done properly. It is far more resistant to scratching and burning. It can be carved, repaired, and polished. Scratches and burns can be sanded out. It is available in several finishes and colors from solid to

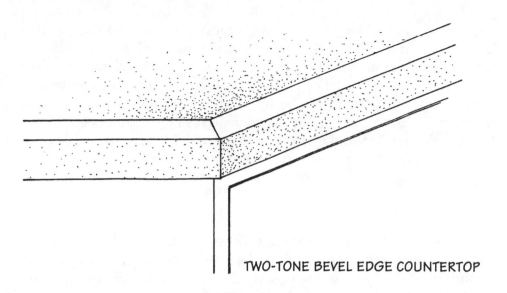

TWO-TONE BEVEL EDGE COUNTERTOP

BEVEL EDGE TOP PROFILE

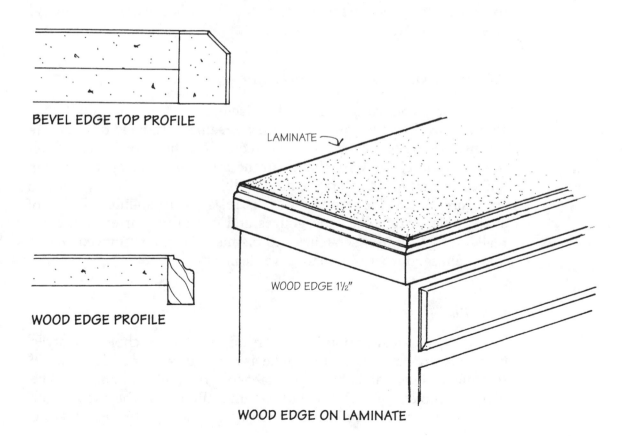

LAMINATE

WOOD EDGE 1½"

WOOD EDGE PROFILE

WOOD EDGE ON LAMINATE

multicolor (much like granite). The biggest advantage to this product is that you can actually have your sink, drainboard, backsplash and wall surfaces made from the same material creating a seamless, one-piece unit. There are no cracks or crevices for water or dirt to creep into. It is extremely durable and will easily handle the toughest cleanup. Cost is generally three to five times as expensive as laminate (depending on the complexity). Use only an authorized dealer in order to avoid negating the warranty and to ensure a proper installation.

Metal Laminate

This is made from the same process as other laminates, but instead of using a melamine surface, they use a thin layer of metal, such as aluminum, copper, and stainless. They offer a variety of styles including brushed, brite, matte, and textured.

Care. As with all metals, they do leave fingerprints and are subject to scratching. Deep cuts will penetrate the surface, exposing the base layer of substrate.

Thermoplastics and Solid-Surfacing Veneers

These are variations on solid surfacing, and they have similar visual and performance qualities with the same advantage of seamless installation. Since they are available in only half the colors and patterns of solid surfaces, they are about half the cost. Also, the intricate inlays and combinations of colors are not possible.

Care. Since it is nonporous, it is resistant to staining. But it is susceptible to scratches and gouges and can burn if it encounters a really hot pot. Minor scratches and blemishes can be removed with a fine-grain sandpaper or nonabrasive cleanser.

Ceramic Tile

This is a great option because it offers so many choices of style, color, texture, and price. You can be as creative as your budget allows. It is highly serviceable, being resistant to scratching and burning. The biggest problem historically with ceramic tile is not the tile but the grout, which cracks, stains, and eventually disintegrates. *Epoxy* grout has eliminated many of the old problems. I do not recommend unglazed tile because it will stain. Also, recognize that if something breakable is dropped onto tile, it will break. For this reason, I often choose ceramic tile in combination with another material, using the tile for areas such as islands and backsplashes. To create a custom look inexpensively, try hand painting several stock field tiles to match your kitchen wallpaper or fabric and then incorporate them into your backsplash.

Care. Mild nonabrasive cleanser, soap, and water. Grout should be occasionally scrubbed clean with bleach and resealed with latex grout sealer.

Natural Stones—Marble and Granite

These are the two most popular types. *Marble* is elegant but very porous. As a result, it is easily scratched and stained. I recommend filling and sealing and periodically resealing marble. Also consider not having it highly polished, since this will only make more obvious the areas that get the most use. It is best used as a countertop for a baking area.

Granite has a multicolored speckled appearance. It is a better choice because it is harder (quartz-based) and resists water, scratches, stains, and hot pans. It should be polished, usually to a high-gloss.

The cost of either is usually six times that of laminate. Since it is a *natural* material, no two pieces are identical. I usually shop around with a client to find a slab with the right look and color. I recommend a thickness of $1\frac{1}{4}$ inches versus $\frac{3}{4}$ inch. It is less prone to cracking. This is particularly true when installation requires it to span an unsupported space of more than twelve inches. Installation is absolutely critical. Be sure you work with someone who specializes in marble and stone and has the facility to properly cut, polish, shape, and install. I have seen too many jobs that were poorly executed in an attempt to cut cost. It is really sad to see an expensive countertop that looks homemade instead of handmade.

Care. Maintaining marble can be difficult. You must take precautions to protect it from staining. Even water will leave a ring. Stains can be treated with commercial products. For coffee and tea stains use white paper towels soaked with 20 percent peroxide. Citric acid etching can sometimes be removed by using a commercial marble polish.

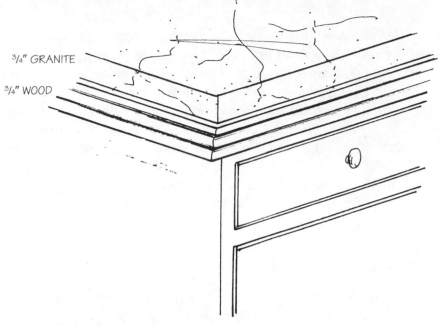

¾" GRANITE

¾" WOOD

GRANITE TOP WITH WOOD EDGE

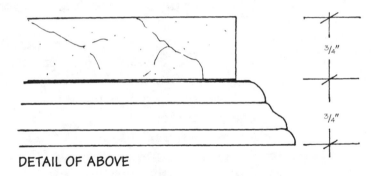

¾"

¾"

DETAIL OF ABOVE

New solvent-based penetrating sealers are best. Granite is far easier to live with. Use a damp cloth and when necessary, a commercial cleaner.

Butcher Block

This can be made of oak or maple and is generally not used for an entire kitchen countertop. Instead, it is used for chopping or as a decorative element within the room. It is made of several pieces of wood that are glued together under pressure. And with time, and improper care, the seams can come unglued. Its natural characteristic is to darken with age. It will show scratches and burns, which gives it character.

Care. If it's unsealed, it needs to be seasoned with mineral oil before use and occasionally thereafter to prevent staining. I do not recommend sealing with polyurethane or varnish, because it does not hold up well. The biggest concern with butcher block is bacteria contamination. Always clean it thoroughly after use with a damp cloth and detergent, rinsing and drying well.

Stainless Steel

This was popular in the 1950s and is gaining popularity again. It can be custom-fitted and shaped, making it very flexible. Custom fabrication allows for integrating sinks and drainboards, as well as custom edge treatments. It can be incorporated right into the backsplash. You can emboss a pattern, such as diagonal checks, into its surface to create a unique design for a backsplash. It is practical and nearly impervious to staining. Over time it will scratch and its finish will become duller. Consider this to be part of its appeal and character. Cost will depend on how creative you get—generally somewhere between laminate and solid surfacing.

Care. Use dishwashing soap and dry with a towel to prevent streaking. Baking soda can be used for tougher spots. To renew the shine, use a cloth dipped in vinegar or use a specialty stainless-steel cleanser.

Paint

Yes, it is an option. If you have a countertop that is in good shape except for an outdated or ugly color, you can paint it. You can even make it look like granite or marble. Many paint manufacturers now make special primers designed to make paint grip tighter to such surfaces. It is not necessary to sand or rough up the surface of your laminate countertop. Durability will be determined by the kind of topcoat sealer you use, and the number of coats you apply. See a more detailed

Painting your old countertop is a way to create a new and beautiful look inexpensively.

description of how to do this in my first book, *My Name Isn't Martha, But I Can Decorate My Home.*

Care. Use dishwashing soap and a damp sponge.

Words of Wisdom

❀ Before ordering your new all-white kitchen, be sure you are prepared to take the time required to keep it looking clean.

❀ If you are having a wood floor installed, have it delivered three days before installation, and put it in the room in which it will be installed. This will allow the wood to adjust to the temperature and moisture of your home.

❀ Allow twenty-four to forty-eight hours curing time after installation of a floor before putting any heavy furniture or appliances on it.

❀ Be sure to install a fire extinguisher near an exit (and away from cooking equipment). Make your kitchen wheelchair friendly.

❀ A kitchen renovation can take anywhere from two weeks to three months, depending on the extent of the renovation.

❀ If you are choosing to work with a professional kitchen designer, get him or her involved in the planning process early. Your floor plan must be set before anything else is decided.

❀ Statistics show that Home Depot dominates the kitchen remodeling market. In 1998, it is expected that they will do 970,541 jobs. This will comprise nearly 21 percent of the total market.

❀ One of the hottest trends in furniture styling is getting strong in the kitchen—the blending of old with new. In the past we ripped out the entire old kitchen and replaced it with shiny new—today, we keep the old and add the new as a complementary statement. It makes for a much more interesting, unique, and cozier style.

❀ Although *islands* have been very popular in kitchens for the past many years, don't feel as though having one is a necessity. Squeezing an island in, just for the sake of an island, is a bad idea. Try using a mobile cart or small freestanding table as an option for a small island in your kitchen. Recently I used a small butcher-block chopping table from a department store (cost $250) as an optional island. It can be moved when you are entertaining to create more space.

❀ One of the smartest ideas I ever came across is by designer Karen Harautuneian, of Los Angeles. She paneled the refrigerator doors with a blackboard. It's the perfect message center!

SMALL FREESTANDING ISLAND

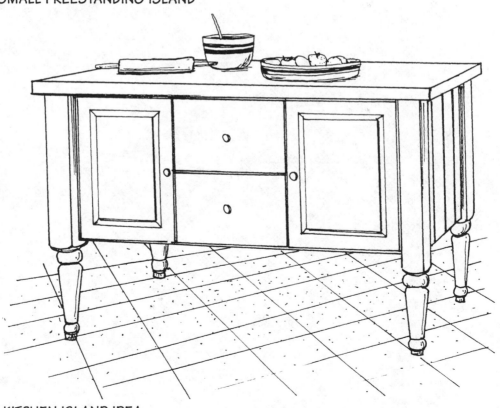

KITCHEN ISLAND IDEA

5

EVERYTHING ELSE, INCLUDING THE KITCHEN SINK

Sinks

You know that phrase "everything but the kitchen sink"? Well I now know why they left out the sink—they have become so complex with built-in rinse baskets, colanders, drainboards, hot water dispensers, soap dispensers, and disposals—that who wants to bother! Then you have the choice of materials: stainless steel, enameled cast iron, and a whole lineup of composite materials. And we haven't even discussed whether you want a single, double, triple, or corner sink. Did you know that the deeper the sink, the more expensive? So where do you begin? First of all, realize that even if you never cook a meal, you will probably use your sink on a daily basis. Choose it for practicality and not for "bells and whistles." The more bells and whistles, the more expensive. Instead, consider how you will use your sink. What are the sizes and kinds of things you will need to wash in it? Will your sixteen-year-old son be cleaning gearshifts in your kitchen sink, and then your daughter laundering her lace camisole by hand? How abusively or gently a sink will be treated will determine what kind of quality and material you should buy. In the past, I have always chosen stainless steel for myself and been extremely pleased. However, for "decorative" purposes, I chose a white composite sink this time. I wanted the sink to blend with my white countertops. Well . . . I have to be honest, it is a lot more work to keep this sink clean than the stainless steel. So in making your own choice, be sure you weigh all the odds and choose what is most important to you.

In most cases, the size and/or choice of sink will be greatly determined by the amount of countertop space you have. If you choose to do a corner sink, be sure you leave enough room on either side to work easily. Also, be sure you don't put the sink too far back from the edge (particularly if you are short like me).

Here are some thoughts to consider when determining size and style.

❀ Is there more than one cook in your house? Then be sure you have room for two at the sink or add a second sink. It may be a bar sink, a prep sink (half the size of a single) with a disposal, or a full-size single sink.

❀ Do you like to soak vegetables while preparing the rest of the meal? You may want to consider a double or triple sink with a disposal in one.

❀ Will your sink be a focal point in the plan? If so, then its appearance is important. Consider some of the more artistic styles, such as kidney shaped, or others that have a rear drain and are sloped to keep them drier (hence cleaner).

❀ Maybe you would prefer a deeper sink. This helps keep the counter cleaner because it contains the water better. Many of the newer models are very shallow. I prefer the deeper. It's much easier for cleaning pots and pans.

MOUNTING

An additional consideration on style includes the method of *mounting*, which includes several options. *Self-rimmed* is basically a drop-in sink with a rimmed edge that sets on top of your counter. *Undermount* is the opposite of self-rimmed; it is installed from underneath the countertop. The countertop edge must be specifically finished to accommodate this type of installation. (This is not conducive to a laminate countertop.) *Integrated* is usually used with a solid-surface material. The entire countertop and sink are one unit. This is great for cleaning because there are no crevices for dirt. A *tile-edged* sink is one specifically designed to work with tile. It is set into the tile. It has square finished corners that get set into the grout just like the tile.

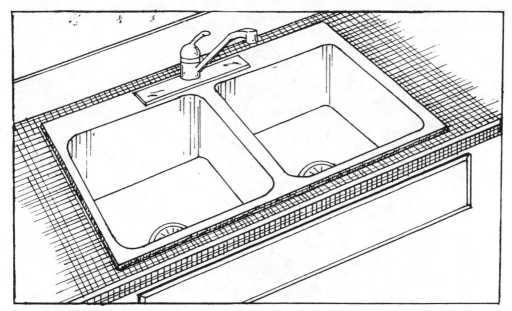

SELF-RIMMING SINK

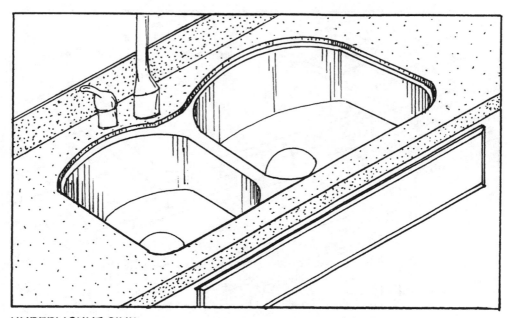

UNDERMOUNT SINK

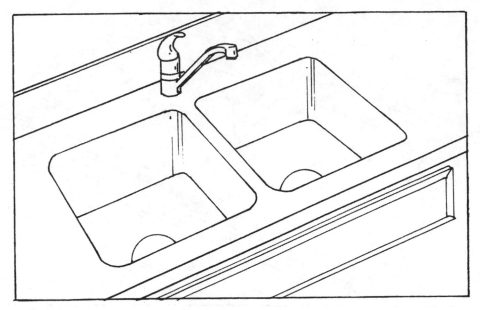

UNDERMOUNT SINK

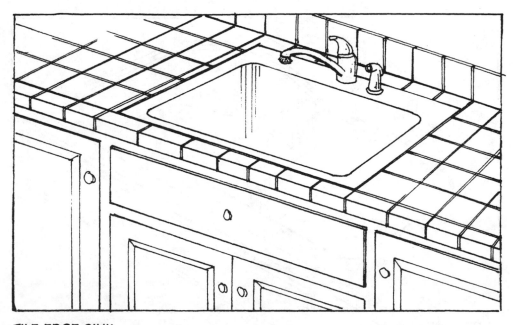

TILE-EDGE SINK

SHAPES

Sinks come in a variety of shapes. Here are the basic options available.

❀ *Single.* If you have limited space, I recommend a large single sink instead of two smaller doubles. In a small space, small double sinks are too small for most chores. I also recommend that you get a fairly deep sink, to make it more flexible.

❀ *Double.* This is the most popular style. My preference is an asymmetrical double—with one sink larger than the other. The smaller one being similar to a prep sink with a disposal. That way you have one larger sink for the bigger jobs with the advantage of a second sink.

❀ *Triple.* Usually the two outer sinks are the same size with a smaller prep-size sink in the center with a disposal. The biggest problem with these is usually they are not deep enough for me.

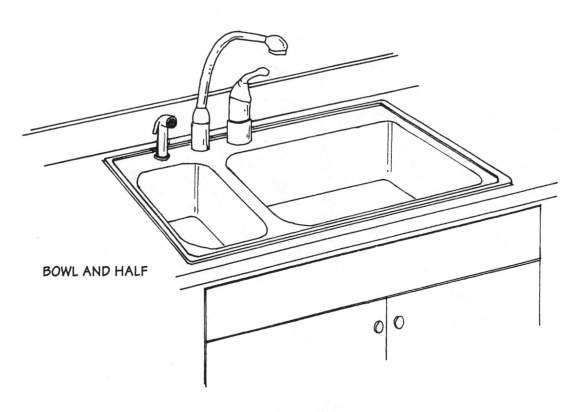

BOWL AND HALF

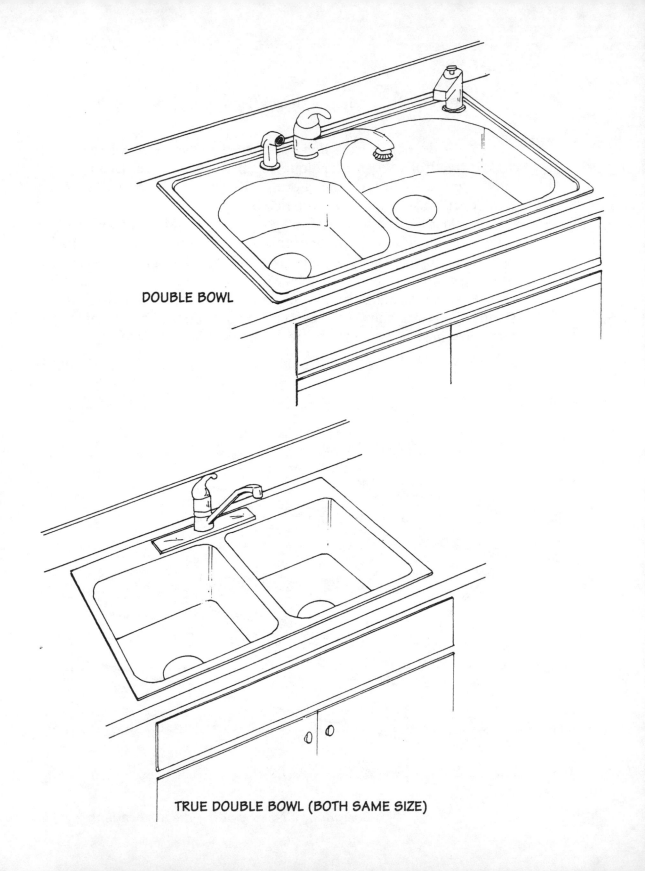

DOUBLE BOWL

TRUE DOUBLE BOWL (BOTH SAME SIZE)

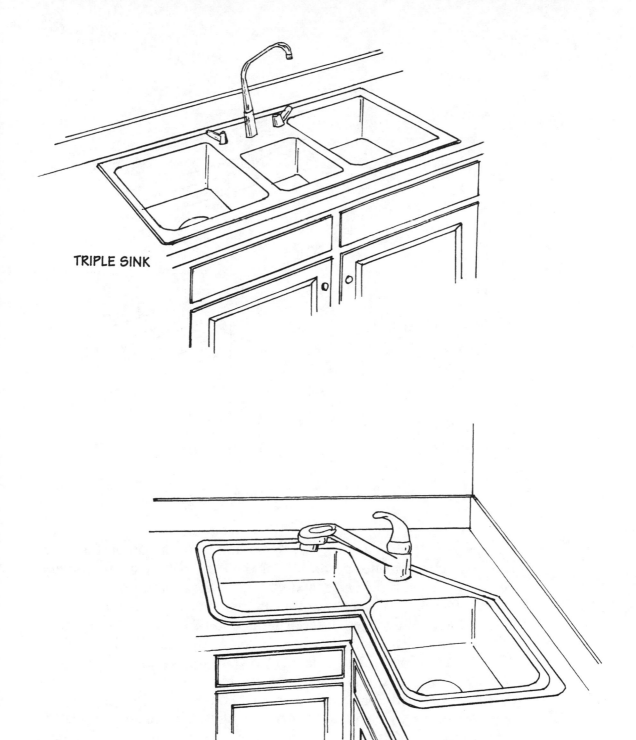

TRIPLE SINK

CORNER SINK

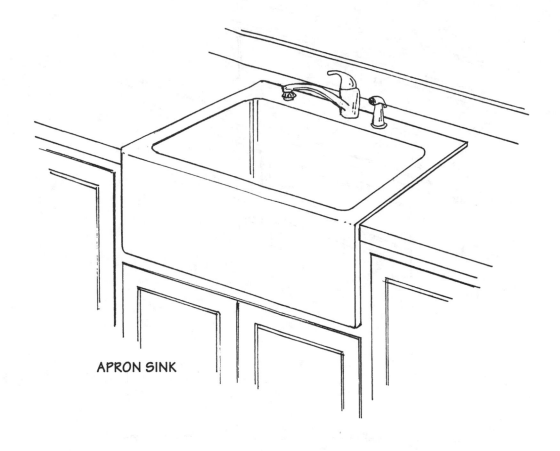

APRON SINK

❧ *Corner.* Most any *single* sink can be set into a corner. But there is a double sink made specifically for the corner. It is set at a ninety-degree angle, with the faucet and sprayer in the center.

❧ *Round.* This is not one of the more popular styles. It is awkward and inefficient. It has less interior room and takes up a lot of space. If you like the way it looks, use it as a second sink, that is, a bar sink.

❧ *Apron fronted.* This is similar to your great-grandma's old sink. Actually that is the intention, a modernized version. It makes a wonderful complement for a country- or farmhouse-style kitchen. They are usually oversize, which makes them wonderful for cleaning up large-size equipment. Often they have a number of options fitted for them such as a cutting board or sink basket.

MATERIAL

Now that you've decided on shape and size, it's time to choose *material*. Here are the basics:

Stainless Steel

This is still the most versatile of all materials. It will work well in almost any style of kitchen. The price and durability is determined by the grade or thickness of the steel. Thickness is defined by gauge. The thicker the sink the lower the number. For example, 22-gauge stainless is pretty thin, which makes it noisier because it creates a tinny sound when it comes in contact with another surface. It is also harder to clean. A 20-gauge stainless is a nice weight. Of course, 18-gauge is even better. The higher the percentage of nickel, the better the sink. The best grade has 10 percent nickel. Usually there are two finish choices available: satin (less shiny) and mirror (very shiny). I recommend the satin. It requires a lot less drying and polishing to keep it looking nice. Price range: single $200 to $460; double $300 to $600.

Care. Usually you only need to use a nonabrasive cleanser. If you have the mirror surface, it may require polishing with an abrasive pad. If so, remember to work only in the direction of the original brushmarks. On occasion, I have used a rust-remover cleanser on scratches that have penetrated the surface and begun to rust. I recommend buying at least one grade above *builder* grade (bottom of the line).

Solid Surface

This is the same material as solid-surface countertops, discussed above. It is great looking and very durable. It is not as easy to clean as stainless steel. In a lot of ways, it is similar to cleaning the old cast-iron enamel sinks. The color is uniform throughout, so scratches generally are camouflaged. I definitely recommend choosing the variegated styles versus solid color because it will show staining and scratching much less. The variegated will cost about 25 percent more. Price range: single $350 on up; double $400 on up.

Care. Use an abrasive cleaner such as Ajax, Comet, or Bar Keepers Friend. Use an abrasive pad such as Scotch-Brite. For more stubborn stains, fill the sink about one-quarter full with a fifty-fifty solution of bleach and water. Allow to soak ten to fifteen minutes. Then drain and rinse well. *White* automotive rubbing compound may be used to remove exceptionally stubborn marks or discoloration.

Enameled Cast Iron

This was a standard just a generation ago. It has been made popular again because of the availability of many different colors. But it still has the same basic properties that Grandma's had. If the enamel is chipped, the iron underneath will rust. Basically price will be determined by how thick the enamel. Thicker is better.

Care. Because of the polished patina (surface), I only recommend a nonabrasive cleaner. DO NOT attack it the way Grandma did with a vengeance and gritty cleanser. Price range: single $125 to $250; double $150 to $400.

Composite Materials

Basically these are made of chipped natural components such as granite or quartz that are combined with a resin (glue). They can be molded into any shape or size and polished to either a matte or glossy finish. They are scratch resistant and camouflage dents and chips well, since the color is continuous throughout. Price range: single $200 to $400; double $300 to $500.

Care. Use only a nonabrasive cleanser. For really tough stains, try an abrasive pad with a nonabrasive cleanser.

An installation note on sinks: Unless you're a plumber, it's one of those jobs that's best left to a pro. A plumber should install your sink, disposal, and dishwasher. Also, your countertop should be installed first.

OPTIONS

Today there is a smorgasbord of optional features for your sink and/or faucet. Be smart. Do not let a fancy salesperson talk you into things you do not need. Most budgets get out of control with all the "little" extras. A few dollars here and a few dollars there all add up to a lot of dollars in the end. Here are the basics with some guidance as to their cost and effectiveness.

❀ *Hot water dispenser.* These were very popular a few years ago. They provide instant hot water (180 to 190 degrees). They are great for such things as instant soup, hot tea, etc. But honestly, unless you are also adding a water purifier, it's probably just as easy to use your microwave. Price range: $275 on up. (I'd rather buy a new dress!)

❀ *Water purifier.* A built-in filter system can provide purer, better-tasting water. If you purchase bottled water, then this is a smart option. It is convenient and cost effective over the long term. Unfortunately, oftentimes there is not enough room under the sink for installation. If you want a water purifier, plan ahead. Most filters improve only the taste and/or odor of water, particularly chlorine. Price range: $170 on up. If you want more, you will need to invest in a water-treatment product. Most of these are designed to address specific pollutants, such as those that are organic, inorganic, radioactive chemicals, and microbiological. The best way to know what you've got is to contact your water supplier or health department and request copies of water-treatment reports for your area. You can also test the water yourself. Contact a national laboratory such as National Testing Labs (800-458-3330) for a home test kit. Price range: $400 to $600. If you have hard water (I do), then you will need a water softener. I am not a proponent of these "electrically charged softeners" that have recently been advertised. I tried one for eighteen months, and it did not work. My faucets were a mess. I now have a standard salt/nonsalt water system water softener. Price range: $400 to $1,600.

❀ *Cold water dispenser.* This is the opposite of the hot water dispenser. Again, I suggest this be combined with a purifier. Price range: $500 on up.

❀ *Sprayer.* These are almost always now part of the faucet package. I definitely recommend having one.

❀ *Soap dispenser.* The goal here is to keep things clean and simple by eliminating the need for soap and/or lotion bottles on your sink top. But they do require filling and cleaning, which often gets neglected. Be realistic about your lifestyle. I chose to purchase an old antique bottle to keep my dish soap. It's pretty and very convenient. Price range: $30 to $95.

❀ *Drain baskets.* This is the updated version of the dish drainer basket. The only difference: This one is custom made to fit the specific size and shape of your sink. Price range: $15 to $45.

❀ *Cutting board.* This too is a custom-fit piece. And as is often the case with custom, it can be expensive. Price range: $30 to $150. (How about season sports tickets instead!)

❀ *Air gaps.* This is really more of a technical option than anything else. Basically, it's a hole for water to flow through, which drains directly into your sink, thereby preventing your dishwasher from overflowing. Some localities require them by law. Price range: $10 to $40.

❀ *Remote drain control.* No, it doesn't operate the TV or VCR! But it does operate the drain of your sink from the upper level. That way you can keep your hands out of dirty water. Price range: $40 to $70.

Faucets

And just when you thought you finally had the sink figured out, you realize you need a faucet! Of course, they too have all kinds of bells and whistles to choose from. The first thing to be sure of is that your sink holes match up with the number of holes required for your faucet unit. For example, determine whether you have a two-hole versus three-hole installation. It is possible to cover an additional hole with a "lid" or hole cover. Be sure you know in advance, so you're not disappointed. Let's start with the basic standard—hot and cold running water.

SPOUTS

❀ *Standard.* Generally this refers to the two-handled version with a spout of 8 to 10 inches long. An updated version would have a spout 12 to 14 inches long, making it easier to reach corners and get large pots under it. The next upgrade would include a "rise" spout that pulls up to give more clearance.

❀ *Waterfall.* This refers to the shape of the spout. The regular spout has an elongated shape that extends forward. The waterfall is higher (10 to 11 inches) and rounded to allow for more clearance. It usually swings 360 degrees. Waterfall spouts are usually available with both two-handled and single-handled units.

❀ *Gooseneck or high arc.* This is an extreme version of the waterfall. It is specifically designed for smaller sinks. As a result, if used on a larger one, it may not reach into the sink but instead splash the side.

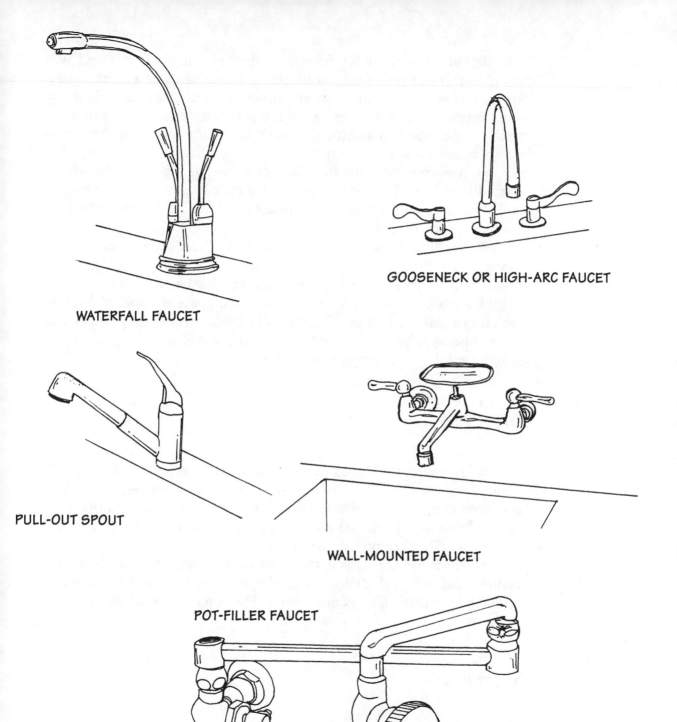

WATERFALL FAUCET

GOOSENECK OR HIGH-ARC FAUCET

PULL-OUT SPOUT

WALL-MOUNTED FAUCET

POT-FILLER FAUCET

✤ *Pullout*. This is my favorite. I first saw it on a boat and was delighted when they became available for home use. It is a combination spout and sprayer in one. The spout actually pulls out of its housing and becomes a hose sprayer. Often it has two different flow options as well: regular and diffused (dual action). Generally this is only available on a single-handled model unit.

✤ *Wall-mounted*. This item is pretty rare. It often is seen with a soap dish attached at the top. If you are choosing to do a farmhouse-style kitchen, you may want to consider this style for its authenticity—but not its practicality.

✤ *Pot filler*. This is something new. A faucet at your stove! It is designed for filling large pots at the stove, rather than having to lug them from the sink. It is wall mounted and folds back when not in use. I think it's a great idea, however, it does require additional plumbing, which can make it cost prohibitive. Oh—by the way—the pot will still be very heavy when you remove it from the stove! Buy this only if you really need it. Price Range: $200 to $500.

HANDLES

✤ *Two-handled*. This is the old standard. I recommend using the "wing" or "lever" style. They are easier to operate with wet hands. You can even use the flick of the wrist. The other style often available is "cross handles." These are considered more decorative for older style renditions. They are not as easy to operate.

✤ *Single-handled*. This is the most popular choice today. It is convenient and practical. The size and shape of the handle is often determined by the material it is made from. They include: solid, looped, and sculpted.

MATERIALS

All of the above faucets are available in a variety of finishes and materials. Some use a combination. Here are the basics along with recommended care instructions.

Chrome

This is the one with which we are most familiar. It is still durable, reasonably priced, and elegant. It will fit most decor styles. The thickness of the chrome plating will ultimately determine price and longevity.

Care. A soft cloth with water or nonabrasive cleanser. Dry well to keep shiny and streak free. Many brands now offer a new protective finish that keeps its shine longer.

Brass

I love the look of brass, but not the impracticality of brass. No matter what kind of protective finish is used, eventually it will tarnish and or chip. It must be kept as dry as possible. And the more chemicals it comes in contact with the worse things will be. It is available in polished, matte, antique, and brushed finishes.

Care. Never use an abrasive cleanser. Use only a soft cloth and water.

Pewter and Nickel

These are the more decorative and expensive options instead of chrome. They have a richer luster. They are generally more durable than brass but not as durable as chrome.

Care. Soft cloth with water.

Colored Epoxy

One of the most popular styles, it is available in many different colors to coordinate with your sink or countertop. It is a tinted coating that is baked on. It is low maintenance and pretty.

Care. Soft cloth with water.

Price ranges for faucet units: a basic unit—$60; midrange—$120; top-of-the-line—$250 on up.

There are some new and unusual terms being used in faucet jargon today. Here are a few just to get you up to speed.

❀ Ceramic disc valves. (Sounds like a car part!) This is a hardened ceramic that is more durable than standard valves. Its manufacturers claim it cannot wear out and therefore won't leak. They are maintenance free and unaffected by temperature changes. Even debris should not bother this valve. It makes a good choice, particularly if you have hard water.

❀ NuSeal valves (American Standard brand name). This is an all-brass compression valve instead of the standard rubber washer. NuSeal uses a plunger-type diaphragm that lifts and lowers as you turn the faucet on and off.

❀ Hot-limit safety stop. This is a setting that allows you to restrict how far the handle can be pushed toward hot. It is a great idea for children or the elderly.

❀ Pressure-balancing valves. This is another term for antiscald valves. Basically this eliminates a change in water temperature when a toilet is flushed or a dishwasher or washing machine is running. It is most often used in a shower installation.

❀ Thermostatic valves. This is a way of selecting, via an LED display, your favorite temperature.

❀ Low-lead manufacturing. As a result of our awareness of the dangers of lead, a whole new sophisticated manufacturing process has been developed. This method utilizes metal molds instead of sand molds for manufacturing faucets. This allows for the use of less lead and a quicker solidification of the unit. As a result, it can reduce the amount of lead to 1.5 percent (less than one-fifth the allowable EPA lead level). American Standard brand is a leader in this process.

Appliances

Take a deep breath—this is a very technologically sophisticated subject. The availability of so many features—such as dual-fuel, air-circulating, environmentally friendly, energy-efficient, and degree-specific temperature control—will make your head swim. So let's first explain what exactly constitutes an appliance. For the sake of discussion, I will define the following as *appliances*: food-waste (garbage) disposers, ranges (stoves) and cooktops, ovens, microwave ovens, refrigerators, freezer units, wine coolers, dishwashers, clothes washers and dryers, ventilation systems, ice makers, and warming drawers. (I hope I've covered all of them!)

The next thing you need to be aware of is that there are now two different categories of all the above appliances being used in homes today—commercial and residential. The most often used commercial-grade appliance in the home is the stove. The refrigerator is second. I will, for the most part, stick with residential. For more information on commercial, you will need to go directly to a commercial kitchen specialty supplier. It is important to recognize that most of the commercially equipped units require commercial-style ventilation and Installation, which can be costly and difficult to implement in your home.

For the most part, I recommend spending a little more on an appliance. That way, you can expect to get many years of good service from it. The extra expense is worth it to have a dependable, trouble-free unit. If you have ever had the experience (and I have) of losing all of your Thanksgiving dinner (on Thanksgiving morning) to a refrigerator gone haywire, you know what I mean.

Appliances are sold in a lot of different places—kitchen showrooms, appliance showrooms, discount stores, home centers, department stores, some interior designers, and even your local gas utility provider. Here in Pennsylvania, UGI (our local gas supplier) has showrooms of the most incredible gas-operated appliances. Next to price and features, the next two most important things to consider when buying an appliance are: warranty and installation.

A great way to get a bargain on an appliance is to shop for showroom floor samples, dent sales (slightly damaged exteriors), and closeouts. Just be sure to inquire as to warranty availability and inclusions.

I suggest that you let the professionals handle any dangerous animal such as gas, electric, or water. Yes, water can be dangerous if it ends up where it is not supposed to be—like flooding my home. Besides, it just doesn't make sense to spend good money buying something new and then wasting it by doing a haphazard installation. In some cases, you can actually invalidate a warranty if the unit is not properly installed.

Now that you've listened to my little "opinion," let's look at the individual appliances and their options.

Ovens and Ranges and Cooktops

First of all, let's clarify the difference between a range and a cooktop. A range is a unit that usually includes an oven beneath. A cooktop is just that, a cook unit (burners) that sits on top (installed into) of your countertop—no oven included.

Ovens

Some of the things to consider when purchasing an oven are: color, style, energy source (gas or electric), and special features and/or options. Energy sources are often determined by availability. I grew up cooking with gas, but when I moved to another state, it wasn't available. I cannot tell you how difficult I found it to switch to electric cooking. It took years before I could fry an egg without breaking it! When I moved to Pennsylvania, gas was again available to me. By that time, I had been cooking with electric for so long, that I didn't want to repeat the learning process in reverse—I stuck with electric.

You also need to consider your individual needs and lifestyle. Ovens are available not only as part of a range but also as independent units such as wall ovens, undercounter ovens (also referred to as slide-in), single ovens, and double ovens. This gives more flexibility in the placement of ovens. Since food cooking in the oven typically requires less attention, you can choose to locate your oven away from the busier centers of activity in your kitchen. Wall ovens can also be located at eye level, making it easier to get things in and out of them. An undercounter oven is a great way to save space while making it possible to use your cooktop without disturbing the cookie baker in the house.

The three basic considerations are: self-cleaning, convection, and dual-fuel systems. (There is one manufacturer offering something new—steam circulating. See below.) Self-cleaning has nearly become a standard. It just makes life so much easier. *Manual-clean* is still available and is less expensive. Today's conventional oven cleaner products have improved, making it a little easier to maintain a manual-clean oven. But they still require some elbow grease. A manual-clean double wall oven starts at $750. If one of them has the self-cleaning option, expect to pay $1,000.

There are now two types of self-cleaning ovens available: self-clean and continuous-clean. The *self-clean* only works if you use it.

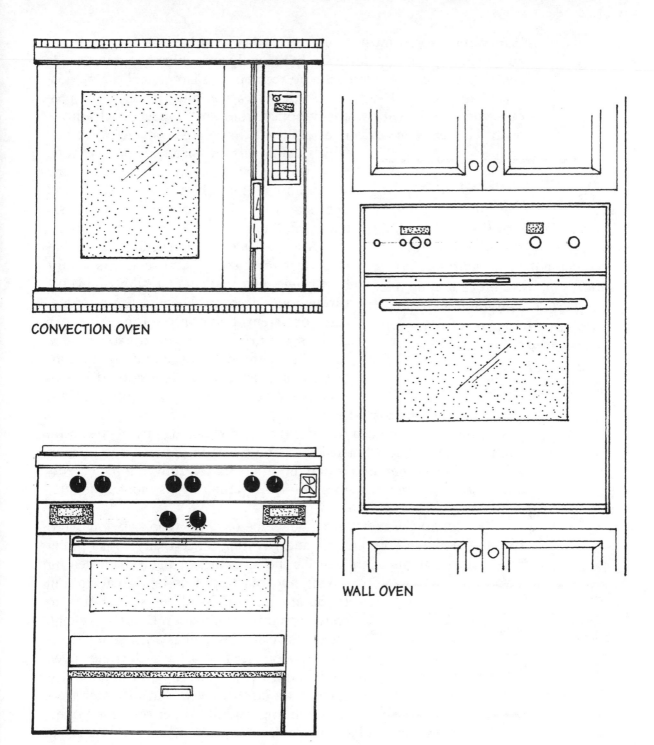

CONVECTION OVEN

WALL OVEN

RUSSEL RANGE

Which means you must plan a time when you have several hours to allow it to operate. Unfortunately I usually remember the oven needs cleaning on the afternoon I intend to prepare a large meal for friends. Then I'm faced with not enough time to accomplish the cleaning process and end up with a smelly house from an "in-need-of-cleaning" oven. The *continuous-clean* oven slowly disintegrates spillage over time, which eliminates the need for preplanning. However, I don't think it is as effective as the self-cleaning method. A self-cleaning wall oven starts at $650. A double wall oven costs about $2,000.

Steam circulating is a new technique that has just been introduced for residential ovens that keeps meat moist without adding fat and gives breads a crisp crust. I know of only one manufacturer currently offering steam circulating, and that's Russell Range.

A *convection* oven is a specialty oven that is perfect for baking because it has a fan/heating element that circulates hot air evenly throughout the oven, keeping the temperature consistent. Food on the top rack cooks at the same rate as items on the lower rack. In addition, food cooks about 30 percent faster. As a result there is less tendency to overcook food, causing it to be too dry. Convection wall ovens cost $1,300.

Something you need to consider when choosing any of the above ovens is the interior dimension of the oven. In general, the interior space is smaller than in our grandmothers' days. Convection elements and/or the required insulation for self-cleaning ovens reduce the interior space.

Dual-fuel systems refers to a new type of range (cooktop and oven) that lets you use both electric and gas fuel. Since many cooks prefer the control that gas offers but like the reliability of an electric oven for baking and roasting, this is a great option.

Be sure to measure the size of your favorite oven ware to be sure it will fit into your new oven.

If you are currently cooking with all-gas, to convert to this combination is fairly easy and inexpensive. Viking makes an amazing dual-fuel model. It has a self-cleaning oven equipped with 1,000 to 15,000 Btu gas burners and an electric oven that offers convection, broiling, and five other cooking modes. It is 36 inches wide and 24 inches deep. It also has an amazing price: $4,675 to $5,550, depending on the finish and configuration.

Many ovens now have *digital* temperature displays, which helps you monitor oven temperature more accurately. Another option you may choose is *delayed-time* cooking. This basically allows you to tell the oven when it should start cooking. Personally, I have never used this feature, even though I have it on my oven. It's just one of those things I'm not comfortable with. I don't like the idea of my oven being on if I'm not home.

Warming Ovens. These are actually an old idea that has come back again. It is simply a pullout oven that keeps food warm at low settings. They are ideal when preparing for a large crowd. It's a good place to keep the turkey warm while you mash the potatoes. It is possible to keep food in a warming oven for as long as two hours.

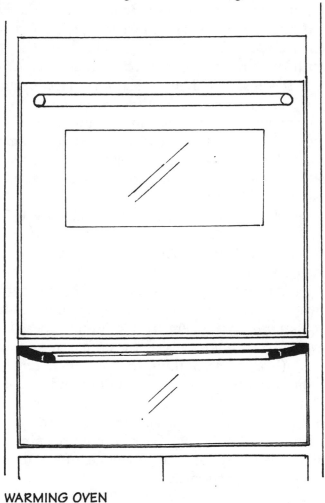

WARMING OVEN

Ranges. Just as with ovens, you will first need to decide which form of fuel you prefer. The advantage to gas is that it responds immediately. If you want hot, it's hot, without your waiting for it to warm up. You also can adjust the temperature down immediately, which is great for a pot about to blow! However, many of the new electric burners now offer a similar control-and-quick-response unit. On the average, electric ranges are slightly less expensive and simpler to install.

As with most appliances, there are some standard sizes to choose from. The original standard size for a residential model was 30 inches. If you are planning on replacing an old unit without making any other changes, this is the size to buy. I prefer to have a 36-inch range because it provides more room and is easier to manage with several things cooking at once. You may also, depending on the manufacturer, find *staggered* placement of burners or even a fifth burner available on the 36-inch size.

With the advent of commercial ranges' popularity in the home, many of the residential manufacturers have begun making several larger sizes. Professional-quality ranges—40, 45, 48, or even 60 inches—are available in heavy-duty stainless steel. They usually have more powerful burners than a standard residential model. But be aware that you will be giving up a lot of space in your kitchen to accommodate one of these beauties, not to mention how much of your budget will also be gobbled up.

Some of the other new options available today include:

❀ *Reigniting burners.* Will relight a gas flame to the same setting if it's accidentally blown out.

❀ *Heat indicator lights.* Usually available on an electric range, they show when a burner is on, or even if it's still hot although it's off.

❀ *Burner options.* There are a lot of new ideas here—such as griddles, wok rings, simmer plates (to diffuse heat evenly), and "bridge burners." (No, it has nothing to do with telling your boss off!) A bridge burner allows you to convert two individual burners into one larger burner.

❀ *Glass window.* This is almost a standard. It is a window on the oven door that allows you to peak without opening the door.

When shopping for a range you also need to know and consider some technical aspects such as:

❀ *Btu's* (British thermal unit). This is the measure of heat output

KITCHEN WITH COMMERCIAL RANGE

from gas burners. The average is about 9,000 to 12,000 Btu's. Some models now offer a specialized low- or high-rated burner. *Watts* is the comparable measurement for electric burners. The standard average is between 1,000 to 2,600 watts.

❀ *Dual element.* This refers to a specialty electric burner that allows you to use either the entire coil or just the inner part. It may also refer to an oven where the top and bottom elements cycle on and off for more even heat control.

❀ *Electronic ignition.* I'm not referring to your car—electronic ignition is now available for your range! An electric spark is used to ignite the gas instead of a pilot light. No more worry over the pilot light going out and being overcome by gas fumes.

❀ *Sealed burners.* These eliminate the gap between the burner and the drip pan on a gas range or cooktop. *Ceramic Glass* is a cooktop style with the electric elements beneath a smooth glass surface. Be sure to use cleaning products specifically designed for this type of glass. Otherwise, you may find yourself with a mess that is very difficult to clean up.

❀ *Touch pad.* This type of control is operated by the touch of your fingertip. The warmth and pressure activate the control panel.

Of course, these new-fangled options are not without their price. A self-cleaning gas range with electronic ignition, sealed burners, and higher Btu's is about $600, compared to a self-cleaning, electric-coil range, which starts at about $400. A 60-inch-wide commercial-style range can cost $7,500. The ceramic-glass top ranges start at $600, and this doesn't include a convection oven, which can add another $250.

Cooktops

(Most of the information given above for ranges is applicable to cooktops.) This is basically the modular version of the stove. Remember when a stereo included everything you needed? Then came *components,* which meant you had to purchase each part individually. Well, cooking is headed in the same direction. Ultimately, this gives you far more flexibility in design as well as the opportunity to customize your kitchen to your own cooking preferences. Interchangeable cartridges allow you to reconfigure your cooktop at will. You might choose to have two burners to one side and a grill on the other, which allows you to grill indoors year-round. You could also choose to have a *down draft* ventilation system in the center. (I'll give more information on ventilation later in this chapter.) Many people also are choosing to have one burner with 600 to 700 Btu's for low-temperature cooking, and another at 12,000 Btu's for large pasta pots.

Ceramic glass is probably the fastest growing trend in cooktops. It offers sleek design and is easy to clean.

There are two types of quick-response elements available for electric cooktops—*radiant* and *halogen.* Radiant burners use coiled wires or "ribbon elements" to heat up, and they give a red glow when hot. Halogen burners use a halogen bulb. Halogen burners are reputed to be more responsive, but they are also more expensive. Some manufacturers are now offering a combination of both types of burners on one unit. I suggest you try different models and makes before choosing.

Magnetic induction isn't just for Star Trek—you too can cook with induction!

Magnetic induction burners heat by means of an electromagnetic field created between the cooking element and the pot (or pan) being used. The only problem with this method is that it doesn't work on aluminum. Therefore, you must use pans made only of steel, a steel alloy, or iron. The point of all of this is that the only thing getting hot is the cooking utensil. The rest of the cooktop surface remains cool. How popular this will become is anyone's guess.

With all these options and features, don't forget to think about the basics. Here are some tips that can help you get back to reality and make choices based on some necessary facts:

❀ Measure your favorite pots, pans, and oven ware to be sure it will fit and function well with your new unit. If you use a lot of oversize pots, you may want to consider a smooth-top range/cooktop, because it will allow the most flexibility.

❀ Choose a model that fits smoothly. Many ranges and cooktops have a self-edged ridge around them. This is designed to keep spills from leaking onto your countertop.

Choose colors and/or finishes wisely. Be sure that what looks good to you today will still make you happy ten years from now. Realize that even if you are choosing white, it may not match your white cabinets, floor, countertop, wallpaper, or refrigerator. There are hundreds of shades of white. Match carefully. (A word of caution: Most white cabinets will yellow with time.)

❀ Consider a range/cooktop with the controls mounted on the top rather than the front if you have small children around the house.

❀ Check to be sure the cooktop lifts easily for cleaning.

❀ Be sure controls are easy to see and read. Can you tell which one controls which burner?

❀ If buying a gas range/cooktop, be sure the flame has infinite variable control. Some models automatically adjust up or down at intervals, not giving you the kind of control you probably wanted.

❀ The more attachments, bells and whistles, the more potential for service needs and the more to clean. Buy only what you think you will really use.

Microwave Ovens

Microwave ovens have really gone through a transformational period. When they were first introduced in the 1970s, it was thought that they might become the only way to cook and replace the stove and conventional oven altogether. So, for a while, there was a tendency to make them larger—large enough even to cook a turkey—with additional cooking features such as convection temperature cooking.

As things have progressed, the food industry met this challenge with more prepared foods that can be cooked in both conventional and microwave ovens. Ultimately, we all decided that some things—such as baking, roasting, and browning—are just better done the "old-fashioned" conventional way. So, the microwave is just another appliance, relegated to its specific area of expertise. As a result, there is now a micro-microwave oven—smaller, compact, and mounted under a cabinet. The microwave/convection and countertop models are still available, but they are not what most people are buying.

You may also choose a space-saver model that is mounted above your cooktop. This often includes a ventilation system. If you have no other space in your kitchen, then this is an option. Be aware, however, that it can make it difficult to use the back burners of your cooktop, especially if you are using a large pot. If you have children who often use the microwave, I would especially hesitate to have them reaching over the stove to get to the microwave. Many years ago, I caught my bathrobe on fire doing just this.

The most important features to consider in a microwave are: interior capacity, wattage (power—the highest available is 950 watts), and whether or not you want or need a turntable. Other options include preset cooking levels for such things as popcorn, reheating food or beverages, and defrosting.

Most microwave ovens are available in black, white, or chrome. Many of the manufacturers have trim kits available, which make it easier to create a built-in look.

Ventilation Systems

No discussion on cooking could be complete without a conversation on ventilation. For most of us, we usually think in terms of ventilating steam and/or smoke. But there is a lot more to it than that—steam, grease, odors, and gas fumes (if you cook with gas). Not only can they steam up or stink up a room, they can also ruin cabinet finishes, wallpaper, draperies, and countertops. So this is not the place to skimp on price or technique.

There are basically two types of ventilation systems available: *updraft* and *downdraft*. You may have heard of a "third" kind—ductless—but it is also (in my opinion) worthless. It only circulates the bad stuff through a filter and then recirculates it into your room. A ventilation system should be ducted to the outside of your house.

Updraft ventilation uses a blower and a hood. It captures cooking grease and fumes as they rise. This is the type you are probably most familiar with, particularly if you currently have a hood above your stove. This is also the only kind of ventilation system to use if you are planning on having a commercial-style stove.

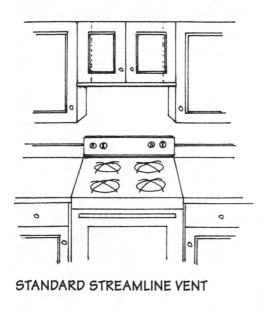

STANDARD STREAMLINE VENT

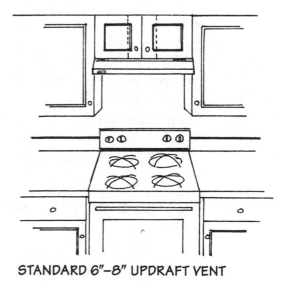

STANDARD 6"–8" UPDRAFT VENT

I have a rather nifty version of the updraft hood. It is compact in appearance—only an inch and a half high—and the rest of the unit is hidden in a cabinet above the stove. The unit operates by pulling forward a sliding glass shelf. This activates the fan and the light. A switch allows you to control the fan power. Obviously, this means that a lot of the upper cabinet is not usable, but it's a lot prettier to look at than a hood.

Another option is to have a beautiful wood hood made to house the standard updraft system. This is a beautiful choice but must be planned for in your budget as well as in your kitchen design.

I recommend getting a fan that is powerful enough to remove the unwanted fumes without scaring you out of the room.

Downdraft ventilation systems are located just above the burners, using pressure to draw hot air and fumes down for removal to the outside. This type of system is effective with low pans or pots but loses some of its effectiveness with taller and/or larger pots. The advantage, however, is that they are less obtrusive. In some cases they are invisible until you need to use them. They pop up with the push of a button. They are usually the best aesthetic choice for an island cooking center.

The technical terms used to define fan power are: *cfm* (cubic feet per minute) and *sones*. *Cfm* rates how much air the unit can vent—a low-end unit will be rated at around 150 cfm; the standard residential unit is rated at 350 cfm; and commercial-style ranges require about 1,200 cfm. *Sones* measure sound. The lower the number, the quieter the fan. A standard residential fan ranges from 4.5 to 7 sones.

In terms of price, almost anything goes. You can purchase an inexpensive (and ineffective) ducted hood unit for as little as $50. But you should expect to pay $250 to $400 for a reasonably effective and quiet unit that is rated for residential use. I do not recommend installing a ventilation system yourself—if not done properly, you can end up with a noisy system that stinks!

FOOD-WASTE DISPOSERS

Speaking of noisy smelling things—this one usually is the worst offender. In addition, it's also scary—at least for me. It took me a long

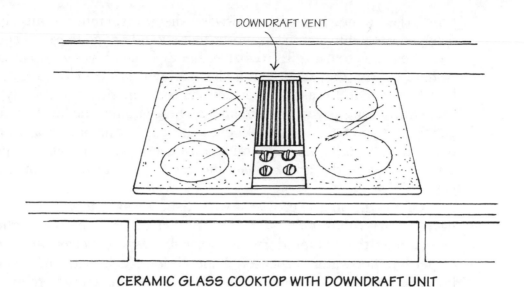

DOWNDRAFT VENT

CERAMIC GLASS COOKTOP WITH DOWNDRAFT UNIT

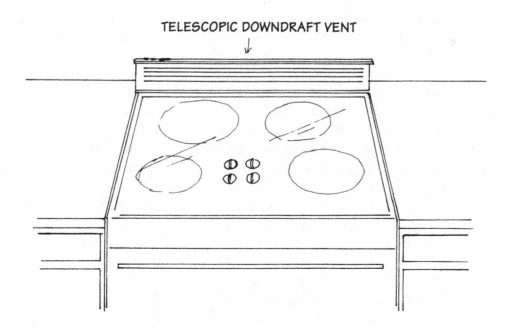

TELESCOPIC DOWNDRAFT VENT

time to get comfortable with food-waste disposers. I must admit, I have dropped a few things into a disposer that should not go there—a spoon and an earring, to name just two. Neither experience was very pleasant. And, of course, they did jam the disposer, which meant I had to learn how to unjam it. In case you haven't had this experience, it's really not too hard, as long as you can remember where you put the little wrench-gizmo that you need to use to loosen the grip of the unit. Many units today actually have an antijamming autoreverse switch that will speed up operation of the grind wheel and break apart anything causing a jam.

One of the biggest problems with disposers is that often you don't have enough room under your sink. Especially with the onslaught of other under-the-sink products being used today—water purifier, soap dispenser, hot-water dispenser, hand lotion dispenser, and sprayer hose. So the first thing you need to determine is how much room you do have. This will help determine what size disposer you can purchase.

The second thing you need to analyze is how often you will actually use a disposer. Think in terms of day-to-day needs. I use mine only once or twice a day, and usually for relatively soft waste, so I don't need the most powerful unit. However, if you are a gourmet cook or entertain a lot of crowds in your home, then you need to get something with a higher HP (horsepower). Disposers range in power from $1/3$ to 1 HP.

There are basically two types of disposers: continuous-feed and batch-feed. *Continuous-feed* disposers operate when you turn the water on and flip a switch. *Batch-feed* disposers start when the stopper is closed and turned; they are also more expensive. I prefer continuous-feed because I feel there is less chance of a child turning it on with their little fingers near the unit. I know of no statistics to back up my feelings, it's just a personal preference.

As I mentioned, noise is a major consideration with a disposer. Never buy a unit without first hearing it in the showroom. Also, the noise level can be affected by the kind of sink you have (stainless steel versus cast iron). I have a composite material sink, and it is a better insulator of noise than stainless steel. However, every time I turn it on my cats take off running for shelter under the bed!

One question that usually comes up is: Can I use a disposer if I have a septic tank? The answer is yes! Experts now agree that if your septic tank meets government standards, food waste will not cause a septic tank to fill more quickly, nor will it harm the system's drain field. I personally lived in a home for twelve years that I built that had both a septic system and a food-waste disposer, and I never had a problem. At that particular period of my life, I did a lot of gourmet cooking and megaentertaining, which put it to the test.

In fact, there is now a disposer available specifically for septic systems. It is made by In-Sink-Erator. It has a replaceable "Bio Charge" cartridge that automatically releases citrus-scented concentrated enzymes into the septic system every time you use it. The cartridge lasts about four months. The enzymes help to continuously break down food and other household waste in the system. The disposer has stainless steel 360-degree swivel impellers and a shredder cutting ring.

DISHWASHERS

In my family, there were six dishwashers: me and my five siblings! It wasn't until we were all grown up and out of the house that Mom decided to purchase an electrically operated dishwasher. (How convenient!) I must admit, some of our best sibling arguments took place at the kitchen sink. "It's your turn to dry." "No, it's *your* turn!" was only matched by, "I did not put the turkey platter in the frig with only one wing on it—it was Lawrence (my brother)!" Anyway, enough reminiscing. As a result, at my first opportunity, I bought a dishwasher.

And fortunately, dishwashers have come a long way in the last few years. They are quieter, more efficient, much better looking, and can hold and clean a lot more than in the past. In addition, they conserve water and energy.

The *quiet* issue is still one of my priority items. Especially with the open-floor-plan style of living. Usually we are in the kitchen or the adjoining family room. A noisy dishwasher (which can take forty-five to sixty minutes to operate) can really be a pain in the ear. The only way to address sound with a dishwasher is with insulation. Obviously, the more insulation, the quieter it will be. I have, on occasion, added extra insulation to the interior spaces surrounding the dishwasher location.

There are a lot of options available on today's models such as: convection drying, touch pad operation, stainless steel interior, built-in water softener, concealed operating controls (to make it more beautiful), delayed start, and decorative front panels. All of which will increase the price in the range of $300 to $1,100. In my opinion, the best development in dishwashers is that you no longer need to rinse the dishes before putting them in. This is due to a combination of things, including better equipment and also better detergents. The other convenience available today is being able to wash good china and crystal in a dishwasher. Specialty racks and different washing levels have made this possible.

If you have hard water, I do recommend using a liquid dishwashing detergent. It works much better.

Under-the-counter dishwashers basically come in two standard sizes: 24 inches and 18 inches. I only recommend the smaller unit if you can't fit the standard 24-inch size in your kitchen. If you do not have room for an under-the-counter model, you may consider a freestanding unit that can be hooked up to the kitchen sink when in use, and rolled away to another space (a closet or pantry) when finished. Usually this convertible/portable style has a removable top and side panels to allow for installation as an under-the-counter style in the future.

REFRIGERATORS

This is one of the most expensive appliances you will purchase. It will also get the most use and abuse, so it's worth spending a little more. It is still possible to buy a refrigerator for as little as $600, but realistically, you will probably spend somewhere in the range of $1,200 to $2,000. One of the reasons for these expensive price tags is that they are now much more energy efficient than they used to be. Another energy issue which has affected their price was the banning of chlorofluorocarbon refrigerants (CFCs), which are harmful to the earth's ozone layer. As a result, manufacturers had to develop new ways of keeping your frig cool. Hence, HFCs—eco-friendly hydrofluorocarbons. Of course, it costs money to develop new stuff.

The first thing you have to decide is what size refrigerator you want. This will depend greatly on how much room you have or can afford in your kitchen for this appliance, and whether you choose a slide-in or a built-in unit. Simultaneously, you will need to determine what kind of a configuration (style) will best suit your needs.

Size

Measuring for a refrigerator requires some forethought. You must think about such things as the swing of the refrigerator door and how much space will be required to accommodate it. If your refrigerator must be in the line of traffic, you will also need to take this into consideration. Be sure to plan for a set-down spot near the frig. This is where you will set down groceries going into the frig, as well as where you will set-down things coming out of the frig. You will need to measure every aspect of the refrigerator. If you are choosing a *slide-in* model, you will need to allow a couple of inches clearance so that the refrigerator door/doors can open fully. If the door cannot open fully, you will not be able to open drawers within the refrigerator. Also, be aware that slide-in models will protrude beyond the depth of the surrounding cabinetry because they are deeper than a standard 24-inch-deep cabinet. One way to eliminate your refrigerator protruding beyond the depth of your cabinetry is to have your carpenter cut a recess into the wall behind it. This creates a little nook that the frig can be pushed into, allowing for a flush fit with the rest of the cabinets. This is what I did in my own kitchen. I then used a standard cabinet above with side panels, which created the look of a built-in refrigerator. Many cabinet manufacturers now have units designed specifically for this. Most of the newer model refrigerators also are designed to *look* built in even though they are not.

If you decide to purchase a built-in refrigerator, be sure to get all the necessary specifications for your cabinet manufacturer. This is critical to getting a good fit.

Built-in units actually have the motor on top instead of in the back. This allows the depth of the refrigerator to be reduced to 24 inches. However, in most cases, the built-in units are often wider than a slide-in unit.

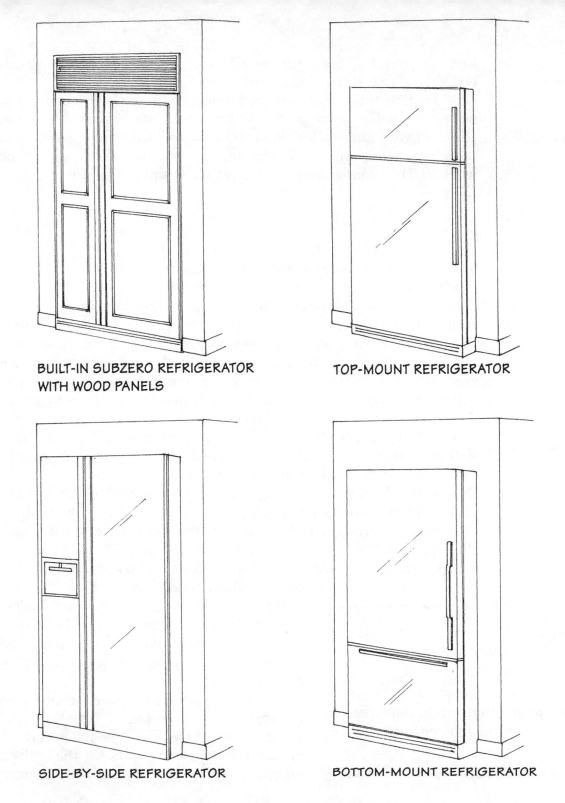

BUILT-IN SUBZERO REFRIGERATOR
WITH WOOD PANELS

TOP-MOUNT REFRIGERATOR

SIDE-BY-SIDE REFRIGERATOR

BOTTOM-MOUNT REFRIGERATOR

Built-in units also generally cost more than a slide-in: $2,400 to $4,200.

Refrigerators are a lot prettier than they used to be. Trim kits, which allow you to customize the appearance, are often available for about $130. A trim kit is designed so you can insert your own decorative panels to match your cabinets. You can also purchase a full kit ($250 to $300), which includes colored acrylic or powder-coated painted panels. I chose a white acrylic panel to match my white kitchen. It's fingerprint proof and very easy to maintain.

Configuration

Configuration is the next major decision to make when buying a new refrigerator. The basic styles available are: top-mount, bottom-mount, and side-by-side (by Sondheim—just kidding). *Top-mounted* styling continues to be one of the most popular. They are what most of us grew up with. The freezer is on top above the refrigerator compartment. The nice thing about top-mount versus side-by-side is the wider interior. The negative side of the top-mount style is that many people feel the freezer is not large enough. Basically you will need to analyze your particular needs and determine how much you generally keep in your freezer versus the refrigerator compartment.

Bottom-mount is exactly the opposite of the top-mount. The freezer section is located on the bottom, usually in a pull-out drawer-style unit. The idea here is if you use the refrigerator compartment more often, then you won't have to stoop over or bend down to get things out of it. However, this is only a good idea if you are tall. I am short and actually find it more of a pain in the neck trying to see what's in back on the top shelf. You will not find a large variety of styles or sizes in the bottom-mount configuration—they are not that popular.

Side-by-side units have really grown in popularity because they have two advantages. The freezer compartment is larger and the double doors are narrower. As I stated earlier, the biggest problem is that the narrower interior makes it difficult to put anything large into it. If you decide on a side-by-side and have a large family, definitely plan to have a second unit somewhere else.

I chose a side-by-side because I definitely use the freezer more than the refrigerator compartment. But . . . every time I entertain, I am frustrated by the narrowness of the refrigerator compartment. I cannot get a large platter in there no matter how I try. Of course, you can always have a second smaller (less expensive) refrigerator in the garage or basement. Actually, many larger families do exactly that. Another consideration when choosing a top-mount is the width of the doors. Depending on the size refrigerator you choose, a door can be as wide as 34 inches. Be sure you can afford this much space in your kitchen.

All of today's units are equipped with a myriad of adjustable shelves, drawers, and other neat storage ideas. Most have temperature control features that are specific to the kind of food you are keeping. Such things as fruit, meat, and vegetables can each be cooled to the perfect degree and humidity. Most refrigerator doors can now accommodate gallon-size jugs and liter-size bottles. You can also choose other convenient features such as automatic ice makers ($60 to $200), through-the-door water dispenser ($200), and built-in drinking-water filtration systems ($100). If you choose to have any of these features, you will need to let your plumber know so he can provide a water hookup for the refrigerator. Maytag even makes a refrigerator that has a special Safekeeper compartment specifically for keeping nonfood items such as film and medicine.

One other refrigerator configuration that has been reintroduced is the *modular unit*. Originally introduced in the 1950s, it is again becoming popular but in a different application. A modular unit is an individual refrigerator unit that can be mounted under your countertop. They are perfect for islands or even near a sink. They are available in drawer and door styles. Some are specifically designed for vegetables and others for wine or beverage coolers. Others are available as freezer units. I think they are wonderful and make a lot of sense. It's a great way to add additional freezer or refrigerator space without taking up a lot of room.

Remember, your kitchen is the heart of your home, and as such it should be designed for you and your family not the neighbors or the trendsetters. Take the time to educate yourself and be comfortable with your choices. Don't be pressured into a quick purchase just because it's a great "deal"—a deal is only a deal if it makes sense for you.

Words of Wisdom

✿ It is illegal in eighteen states to send large appliances to land-fills. Check out laws in your state before you discard an old appliance. Instead, try recycling. Call 800-937-1228 to locate the nearest recycling plant.

✿ NEVER put any of the following down your disposal or you will regret it: rice, pasta, and celery. Any of them will clog, bind, and/or ruin it.

✿ A full freezer is a more efficient freezer.

✿ The bigger the refrigerator, the more electricity it needs for power.

✿ Since dishwasher doors open down, they are more likely to be affected by gravity. This means that the weight of a door panel can be a major issue. Door panels that exceed the recommended weight limit (seventeen pounds) can cause the door hinges to bend and sag, or even break. I suggest checking with your dealer for the availability of heavy-duty springs if you are considering an unusually heavy door panel.

✿ Attach your dishwasher to the underside of your countertop to keep it from traveling across your kitchen when in operation.

6

BATHROOMS

Expectations and Evaluations—Rework or Start from Scratch?

Next to the kitchen, the bathroom is probably the room that is most often renovated in the house. It is also one of the most rewarding remodeling projects from both a financial and an emotional aspect. From a financial perspective, a remodeled bathroom can improve a potential home buyer's attitude toward the home, as well as the amount of money they are willing to spend. It is certainly one of the first rooms a home buyer evaluates when house hunting. Why? Because the bathroom has become our oasis for comfort—our personal spa for finding peace and rejuvenation. A friend of mine believes a bubble bath can cure almost anything that ails you—from a bad day at the office to an ugly fight with your spouse.

I certainly think that the bathroom industry believes this to be the cure of the century, considering the number of new products being introduced almost daily. One manufacturer actually has a self-contained shower unit that includes: a TV, radio, CD player, clock, temperature control, and I think a shower! The cost—$15,000. The same designer also has created the dream bathtub—*J-Sha* (its name)—engineered to simulate a shiatsu fingertip massage. He consulted with masseuses to design the sybaritic tub, with a backrest that has thirty-two microjets, which are activated sequentially to imitate the rolling pressure of an actual massage.

An indication of how obsessed we have become with our bathrooms is that an episode of *Seinfeld* was dedicated to it—Kramer decided to "live" in his shower. He even installed a food-waste disposer in the shower drain so he could prepare meals without leaving the shower!

All of this fantasy bathroom stuff probably started in the Victorian era, and a resurgence occurred in the 1980s with the introduction of the whirlpool tubs. What I find interesting is that now, nearly into the next millennium, we have realized that the Victorians' simple elegance made a lot more sense than our high-tech tubs. In fact, the latest trend in bathrooms are Victorian soaking tubs. These are the old "ball and claw" tubs that are so big and deep (17 to 19 inches) that they can accommodate soaking up to your shoulders. These tubs are wonderful because you can slowly immerse yourself and let go of all your worries. The best part about them is they're quiet. No loud jets, or droning heaters, just the perfect sound of silence. (Of course, this means that the old ball and claw bargain tubs will now be twice as much money.)

I knew early on that most of the clients who insisted on having whirlpool tubs probably would never use them. Why? Because they take forever to fill, by which time the hot water is lukewarm. And turning on the jets (the very loud jets) only causes the water temperature to drop further. By the time the tub is "ready," the phone has rung and it is too late to relax anyway. But in my humble opinion, I'm still not sure we'll manage to find enough time to enjoy Victorian tubs either. The bottom line is: most of us are so busy we barely have enough time to take a quick shower, let alone time to wait for any size tub to fill. But at any rate, these claw tubs are beautiful—almost a work of art. I guess you could consider it a bathroom sculpture. That way you won't have to feel guilty if you buy one and don't find the time to savor it.

The Plan

I'll discuss bathtubs in more detail later in this chapter. For now, let's start with the basics—planning. First (as always) check local building and plumbing codes. Call the building inspector's office to find out which permits are necessary for your project. Then map out the structural components of your house. You need to be sure that there is sufficient reinforcement to hold any heavy bathroom fixtures.

Depending on your particular situation and where you decide to move or replace existing fixtures, you may need to add floor joists, new water supplies, drains, and even vent pipes to accomplish the plan you desire. Plans must be structurally sound, put to paper, and approved by your local codes people *before* you touch anything.

The first step to designing a plan is to locate the plumbing *core* of your home. This is the one central area where most of the plumbing facilities congregate or meet. The most cost-efficient way to add to an existing bathroom's facilities is to work within the existing core. The next consideration should be in keeping fixtures arranged in such a way that you can effectively keep all the pipes installed in a single "wet wall" (a wall dedicated to all the required water pipes), rather than having them spread out, making the project more difficult and more expensive.

Here is a list of basic questions to ask yourself before you begin further planning:

1. What are your needs? Are you happy with the basic layout or does the entire room need to be reorganized? Do you need to add additional square footage to the room? Will this necessitate adding on, or can you borrow space from an adjoining room?

2. Who will be using the bath? If it's a master bath, will children also be using it? Often the children's bathroom doubles as a guest bathroom. Take this into consideration when planning.

3. How much are you willing to spend? Is that figure consistent with the overall value of your home? If not, are you willing to spend it anyway, even if you can't recoup it when you sell your home? Your answer will be determined by how important your bathroom is to you and how long you plan on residing in it. Regardless of what dollar amount you decide, be prepared for it to cost 10 to 20 percent more; it's inevitable.

4. Plan ahead—can you afford to do the entire project at one time? Would it make sense to do it in stages? Perhaps replace fixtures first and cabinetry later.

5. How will you finance this project? This can affect how much you spend. Most people tend to spend more when they finance rather than pay cash. As I mentioned earlier, there is a

new mortgage twist available. It's called the HomeStyle mortgage from Fannie Mae. It is based on a home's appraised value *after* renovations are finished—not its value at the time of application. This allows you to borrow enough money to help finance major repair work. Many lenders offer the HomeStyle mortgage at interest rates that are comparable to standard mortgages. For more information contact Fannie Mae at 800-732-6643.

6. What aspects of this project do you feel qualified to do yourself?

7. Consider obtaining more than one opinion about possible floor plans and ideas. There is always more than one way to arrange a space.

8. Decide who will ultimately manage the project—you or your spouse? There needs to be *one* basic contact person, otherwise there can be confusion, frustration, and chaos.

9. Are you *emotionally* prepared to deal with not having your bathroom available? Your attitude is critical to the outcome of this project.

10. Are you *physically* prepared to deal with not having your bathroom available? What kind of arrangements can be made in the interim to provide for bathroom facilities?

Whether you are adding on a new bath, or remodeling an old one, I suggest making it as large as you can. When I was designing my own home, I stole space from a nearby closet to add additional space to the bathroom. (By the way, this is the closet I designated to my husband!) The rule of thumb says that you need a minimum of 4 by 4 feet for a half bath, and 5 by 7 for a full bath. If you are working with spaces that are this small, it is critical that you clearly understand the minimum clearances required by your local municipality. I have in many cases suggested eliminating a bathtub from a plan in order to provide a larger more luxurious shower, and a more efficient outcome overall. If in fact you nearly never take time to soak in a tub, this is a good way to get a luxurious-feeling bathroom without having to add to the existing space. The drawings on pages 156 and 157 show basic minimum clearances considered to be important to designing a comfortable bathroom.

OLD BATHROOM UPDATED WHILE MAINTAINING
OLD LOOK

NEW CONTEMPORARY BATHROOM

One of the biggest trends, and I believe most practical ideas, is to *compartmentalize* your bathroom. This means that you actually divide up the space to accommodate specific tasks. I think having a *water closet* is one of the most important things you can do. A water closet is a separate compartment for the toilet. That way, two people can be in the bathroom simultaneously while maintaining a modem of privacy, making the morning a lot easier for everyone. In addition, I think the shower should be separated from the vanity area—at least far enough away to prevent the mirror from completely fogging up. If you've traveled, you probably have experienced the luxury of compartmentalization—hotel suites have been designed this way for years. Of course, I think most couples (and particularly the women) would agree the most desirable situation would be to have *two* complete bathrooms—his and hers!

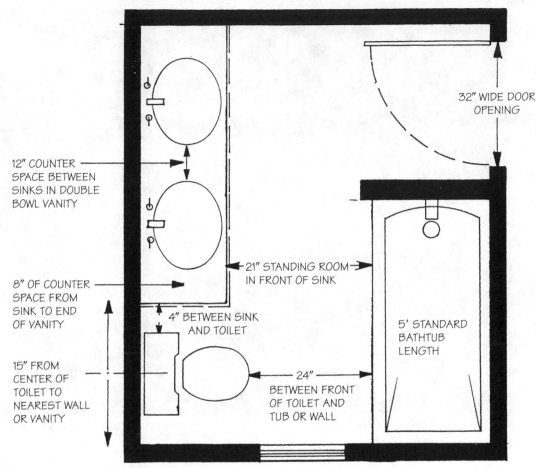

12" COUNTER
SPACE BETWEEN
SINKS IN DOUBLE
BOWL VANITY

8" OF COUNTER
SPACE FROM
SINK TO END
OF VANITY

15" FROM
CENTER OF
TOILET TO
NEAREST WALL
OR VANITY

21" STANDING ROOM
IN FRONT OF SINK

4" BETWEEN SINK
AND TOILET

24"
BETWEEN FRONT
OF TOILET AND
TUB OR WALL

32" WIDE DOOR
OPENING

5' STANDARD
BATHTUB
LENGTH

MINIMUM CLEARANCE FOR FULL BATHROOM

Another growing trend is *universal* design. Universal design incorporates good looks with accessibility for everyone: children, handicapped, and elderly alike. This is becoming more important as the baby boomers mature. Showers that can accommodate a seat or wheelchair, with a barrier-free entrance, will continue to become more important over the next ten years. It only makes sense to give consideration to these issues now, if you are contemplating a renovation project. Even if you don't plan on being in this home when you reach "maturity," it will add to the value of your home's resale price.

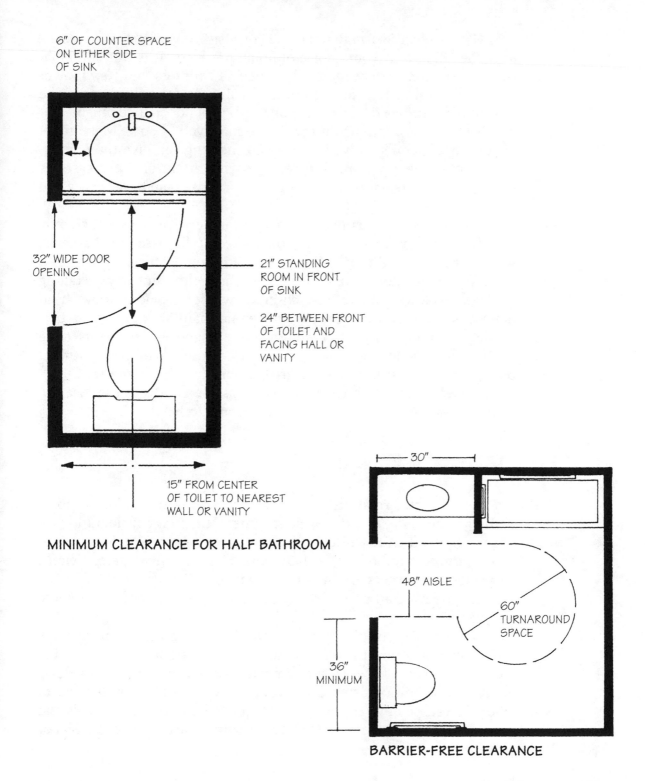

6" OF COUNTER SPACE
ON EITHER SIDE
OF SINK

32" WIDE DOOR
OPENING

21" STANDING
ROOM IN FRONT
OF SINK

24" BETWEEN FRONT
OF TOILET AND
FACING HALL OR
VANITY

15" FROM CENTER
OF TOILET TO NEAREST
WALL OR VANITY

MINIMUM CLEARANCE FOR HALF BATHROOM

30"

48" AISLE

60"
TURNAROUND
SPACE

36"
MINIMUM

BARRIER-FREE CLEARANCE

Bathrooms • 157

Ultimately, the overall impact of your bathroom will most dramatically be affected by your choice of finish products. It is this combination of flooring, surface materials—such as counters and tub/shower surrounds—cabinetry, and colors that will make the difference. The size and sometimes the shape of your bath may not change much—but you have endless options on the final details you choose. Make them count. Small details with tile trims and molding can give the impression of opulence for a small price. By paying attention to accessories such as towel bars, faucets, lighting, and sinks you can have luxury and efficiency all in one.

A bathroom that bathes you in *light* can help wash away a gloomy mood. In addition to artificial lighting, which I discussed in chapter 3, windows and skylights are critical to a bathroom's efficiency. I almost always suggest adding windows nearer the ceiling line. By installing shorter, shallower (15 to 24-inch-high) windows placed above head height, you can add natural daylight while maintaining privacy. In a smaller bath, I often place a window directly above the mirror at the vanity. There is no need for draperies or any other window treatment because privacy is not an issue at this height. (This same technique works well in a walk-in closet as well. Just be sure that the location doesn't allow for direct sunlight to fade your clothes.)

Materials and Budgets

Whether you're starting from scratch with a new addition for a bath or remodeling an existing space, one of the first considerations is usually how much it will cost. Unfortunately, there is no simple answer. Bathroom remodeling is labor intensive—plumbers, electricians, drywallers, carpenters, cabinetmakers, mirror folks, structural support, and designers are all part of the team of experts required to complete the job.

Some trades people charge by the hour, and others by the job. Others will discount if you purchase from them such items as lighting fixtures, plumbing fixtures, and cabinetry. Local rates and the considerations and conditions specific to your job will also affect the cost of labor. The scope of your project may require licensed professionals and most often building permits. This too becomes part of the overall labor cost.

Below I have attempted to give you some guidelines for labor budgets. Recognize that I have priced them individually. It is sometimes possible to get a lower rate from a construction company that can provide all of the services, as opposed to hiring independent specialists for each component.

PLUMBING

- ❀ Replacing a sink: $50 to $200
- ❀ Replacing a toilet: $100 to $200
- ❀ Installing a second sink: $150 to $300
- ❀ Adding a whirlpool: $500 to $1,000
- ❀ Installing a new tub (new location): $2,500 to $3,000
- ❀ Adding a shower stall: $500 to $600
- ❀ Replacing faucets: $75 to $150
- ❀ Replacing a shower valve with a pressure-balanced valve: $300 to $800
- ❀ Adding a shower grab bar: $25 to $75

ELECTRICAL

- ❀ Adding outlets: $50 to $75 each
- ❀ Installing light fixture: $75 to $100
- ❀ Installing light switch: $60 to $70 each

DEMOLITION AND CONSTRUCTION

- ❀ Knocking down 8-by-8 wall: $150 to $900 (non-load-bearing)
- ❀ Knocking down 8-by-8 wall: $300 to $2,000 (load bearing)
- ❀ Removing tile tub enclosure: $300 to $600
- ❀ Installing one-piece tub enclosure: $100 to $200
- ❀ Solid-surface (four-piece) tub enclosure kit: $1,100 (Installing solid-surface tub enclosure: $300 to $400)
- ❀ Fiberglass (five-piece) tub enclosure kit: $1,500 (includes installation)
- ❀ Installing tile tub enclosure: $4 to $8/sq. ft.

FLOORING

- ❀ Replacing subflooring: $200 to $500
- ❀ Improving subflooring with new overlayment: $100 to $200
- ❀ Cost of new resilient flooring: $10 to $40 (installation: $2.50 to $5/sq. ft.)
- ❀ Installation cost of ceramic tile: $4 to $8/sq. ft.

COUNTERTOPS

- ❀ Laminate: $15 to $20 lineal foot
- ❀ Solid surfacing: $50 to $100
- ❀ Ceramic tile: $2.50 to $15/sq. ft.

Now that you have a rough idea of individual costs, here is a guide that will help you determine how to allocate your budget. The National Kitchen and Bath Association survey for 1996 concluded that the average bathroom remodeling project cost $9,300. That figure breaks down to the following percentages.

- ❀ 33 percent on cabinetry
- ❀ 21 percent on labor and installation
- ❀ 11 percent on countertops
- ❀ 8 percent on design
- ❀ 5 percent on miscellaneous extras
- ❀ 4 percent on flooring

I've divided bathroom remodeling projects by cost into several different budgets to give you some idea of what you can expect to get for your money.

$3,000 TO $5,000

Basically you'll get what I call "lipstick and rouge"—a cosmetic makeover with a few extras.

- ❀ New paint and window treatments
- ❀ Base-price new fixtures, i.e., new toilet or sink

- ❀ New resilient flooring
- ❀ Fiberglass tub enclosure or regrout and glaze existing tile enclosure
- ❀ New (stock) vanity with mirror
- ❀ Maybe a skylight or solar tube

$5,000 TO $10,000

If you expect to spend this much money on your bathroom, please get professional design help. This is too much money to spend unwisely. This kind of budget will allow you to upgrade to finer quality products.

- ❀ Midrange price stock cabinets with custom appointments
- ❀ New and/or additional windows
- ❀ Solid-surface countertops
- ❀ Upgraded electrical with new light fixtures
- ❀ Base-grade ceramic tile flooring

$10,000 AND UP

This price range definitely requires the help of a professional designer, and if you decide to add on, then I suggest an architect as well.

- ❀ Add on or borrow space from an adjoining room
- ❀ Custom cabinetry
- ❀ Luxury shower/whirlpool
- ❀ Fancy deluxe model sinks
- ❀ High-end faucets
- ❀ Granite or marble countertops
- ❀ Add a bidet—one-piece toilet
- ❀ Your choice of ceramic tile floor
- ❀ New windows and/or skylights
- ❀ Custom bathroom accessories such as: hooks, door pulls, toilet tissue holders, towel holders, etc.

Sizes Shouldn't Matter

It would be wonderful if we all could afford and had the space to accommodate the luxury of having the ultimate bath of our dreams. But this is not Disneyland, so we must deal with fact and not fantasy. I have attempted to give you guidance based on the basic size of your particular bathroom.

The Small Bathroom

Bathrooms range in size from one extreme to the other. Recently the "enormous" bath has been very popular. Some have lounge chairs, a fireplace, treadmill, television, telephone, security monitor, and are wired for high tech and sound. I would love to know how much time these people actually spend in their bathrooms. Don't get me wrong, I spend my fair share of time in the bathroom too, but . . . there is a limit. If you don't happen to have an 18-by-25-foot bathroom, don't fret. Many homes still have a small or average-size bath. So, I'll start with some suggestions for improving conditions in the smaller bath while creating the illusion of more space.

I am a stickler for organization. I believe in "a place for everything and everything in its place." This is especially important in small spaces—clean up the clutter. Plan to have a lot of hiding spaces. Since most items in the bathroom aren't large, you don't need to have storage cabinets that are 24 inches deep. In most cases, 10 to 18 inches is more than sufficient. Here in Lancaster, Pennsylvania, there are a lot of old historic homes. This means that most bathrooms were retrofitted and as a result are very, very tiny. We have to be especially creative in finding ways to make these near-bathrooms efficient for today's lifestyle. Hanging shelves that include cut-outs for holding such things as hairdryers, curling irons, and the like are very popular. Whenever possible, recess into the walls. It's amazing how much you can accomplish with

I will also give you my standard statement of wisdom: There is no such thing as a perfect house. Nor is a house ever finished—there is always something that needs to be done. If you can accept this fact, this whole process of remodeling will be a lot less stressful.

4 to 6 inches of recessed space. Towel bars, mirrors, and shelving all work within these dimensions.

Build storage into the tub and shower surrounds. A simple shelf on the wall adjoining the tub can be a decorative as well as practical idea. Little nooks and crannies can make a huge difference in small spaces. When I built my shower, I had three triangular tile corner shelves mounted into one corner of the shower stall. The tile man thought I was a little loony until it was finished. It looks and functions very well, giving both my husband and me a place for shampoo, cream rinse, shower gel, and so on without a mess. Don't forget the wall behind the toilet—you can recess cabinetry into the wall or add a free-standing surround.

Consider an interesting freestanding cabinet or add a tall cabinet (much like a kitchen pantry) at one end to create a linen closet. I have added *pier* cabinets to many bathrooms. Pier cabinets sit on top of your countertop—they are only one-third to one-half the depth of the countertop. That way the countertop is still useable, but you've gained a good deal of storage. Another idea borrowed from the kitchen is the tilt-out shelf at the sink. This can be a great place for your toothbrush and toothpaste.

My three favorite and most often used ideas in a standard small bathroom are building: a banjo-top countertop, a cabinet surrounding the toilet, and a shelf above the shower curtain rod. A *banjo-top* countertop is actually an extension of the sink countertop that extends along the wall and over the top of the toilet, in a banjo shape. The banjo shape allows you to narrow the depth of the top at the toilet, to make it still practical. The trick is to allow enough room below the top to still accommodate the removal of the toilet tank lid. You will gain an amazing amount of useable space in a very small amount of room. A *cabinet surround* around the toilet is a fabulous way of gaining cabinet and shelf space. By using the available space above the toilet and going up as high as you desire (up to the ceiling, if you like), you will add quality and practicality to your existing square footage. A *shower curtain shelf* is simply a shelf built above your shower curtain rod. Add an apron front and you will be able not only to gain a decorative shelf but you will also be able to hide an otherwise drab curtain rod in the process.

Using *organizers* inside any storage space is also a great way to get more use out of whatever space you have. Inexpensive wire and basket

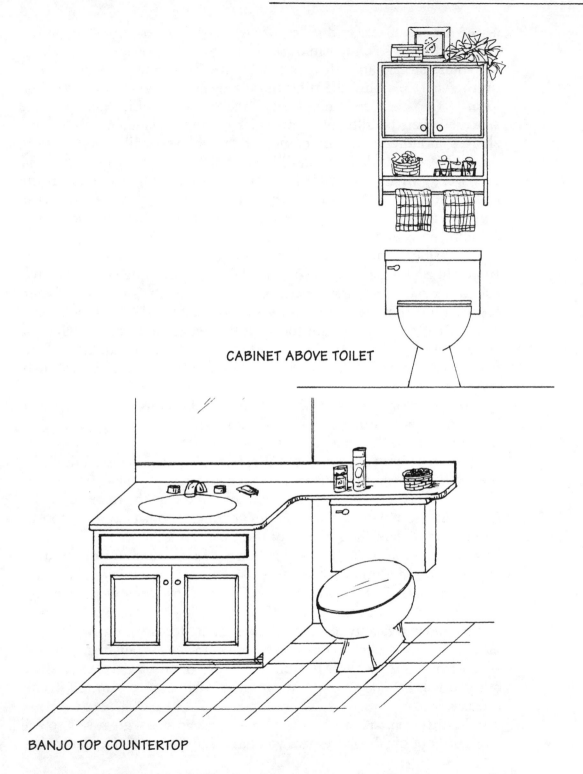

CABINET ABOVE TOILET

BANJO TOP COUNTERTOP

organizers can be purchased almost anywhere—home stores, discount stores, hardware stores, even the pharmacies seem to have them. Many years ago I purchased door-mounted shelving for my bathroom vanity. I have moved them three times over the last sixteen years. They're wonderful!

I always attempt to eliminate sharp corners in small bathrooms. I have learned the hard way how painful it can be to slam your hip against a sharp corner of a vanity. Not only will a rounded or curved shape be kinder and safer, it also creates the illusion of more space. Another tip: If at all possible, avoid *pocket doors* (doors that slide into the wall rather than swing out) because they are not good privacy protectors. By design, you end up with a hollow wall that echoes sound rather than insulates it. The door itself will not have a tight seal, so it too will allow sound to pass through. Pocket doors are also awkward to operate. Additionally, it is very difficult to hang anything on the wall because any nail or screw will interfere with the operation of the door. Be sure this is your only option before making a decision to use a pocket door.

One of the easiest ways to enlarge the appearance of a small bathroom is with the use of mirrors. Often I have mirrored entire walls. Be sure to think about *what* will be reflected in the mirror. In a very elegant model home, with a very large and elegant bathroom, I saw a very big mistake with a mirror—they mirrored the wall adjacent to the bathtub, however they neglected to consider the object that would be reflected into the mirror—the toilet! As a result, it was the first thing you saw when you walked in the room—not very discreet. In fact, it's one of my pet peeves—why do so many builders put the toilet in the most prominent position of the bathroom? Do they really believe it's a throne? Whenever possible, locate the "throne" in a more discreet area of the bathroom.

There are basically two different directions for overall schemes in smaller bathrooms. Some believe the only choice to consider is a "monochromatic" color scheme—one in which everything is a neutral, light color. This can be very elegant and will create an open feeling. But not everyone likes a monochromatic scheme. In that case, I suggest a very creative scheme. By using interesting materials, colors, and design elements, you can create a minibathroom that makes a powerful statement without closing you in. Consider using an old chest of drawers as a vanity, or use an old wash basin as a sink. These kinds of unusual elements will add flavor and style to a small space. I believe this actually works better in a small space than in a large one. In a large space you need to almost "overdo" the interest, which just creates confusion.

It's also important to use "open"-feeling products for areas that will particularly close in the space. The shower is one of those areas. Try using clear or light-colored materials for containing shower spray. Do not use a dark-colored shower curtain in a small bathroom. Hanging, wall-mounted vanities, or the new-again "old-fashioned sink on legs" are also good ways for creating the illusion of more space by allowing more floor to show. Add as much light as you possibly can to a small bathroom. Windows, skylights, false skylights, solar tubes, and additional lighting will make a tremendous difference. Even if your bathroom is without windows, proper lighting can lift the mood of the room and you as well. A small bathroom can feel luxurious with the use of quality products. Choosing rich finish materials such as marble, tile, and finer quality accessories can go a long way in making your bathroom special and not just ordinary.

The important thing is that your bathroom plan functions well without giving you a cramped feeling when you walk into the room. If at all possible, place the largest items on the far end of the bathroom or at least out of your initial line of vision. Give your eyes something pretty to see. It's amazing what a beautiful sink can do for a small space—it's like the crown jewel.

Large Bathrooms

The luxury of the 1990s home is the *large* bathroom. The bathroom industry has jumped at the opportunity to fill the void in the big bath. Dual showerheads, folding seats, pulsating body sprays, soft misting rainbars, drenching waterfall spouts, and therapeutic steam are all available—for a price! Jacuzzi's J-Dream shower, which includes a seat, body spray walls for hydrotherapy, storage space for accessories, and a CD player and stereo, also has a steam generator that lets you convert your shower into a relaxing steam bath. Where will it all stop? Well . . . don't look now, but the government's new regulations mandate that new showerheads allow no more than 2.5 gallons per minute (GPM). Will that put a damper on things? Not exactly . . . there is no mandate as to the *number* of showerheads you may have! Kohler's ten-jet BodySpa, which recirculates water, may be the start of a new wave of products.

A large bathroom allows for the luxury of accommodating different needs and preferences in the bath. This is most often recognized with the design of the vanities. Men for the most part prefer a taller vanity—36 inches—while most women would love a makeup vanity that allows them to sit comfortably.

Of course the bathtub is definitely a focal point in the large bath—whether your choice is a whirlpool or a soaking tub. Most large bathrooms are designed around this sculptured beauty.

Overall, large bathrooms need to be calmer in color palettes, otherwise they become chaotic instead of soothing. As a result, most of the design interest is created by using interesting materials of different textures. Such things as marble, tile, mirror, lacquer, and wood in similar colors can make for a very interesting yet cohesive look. Repetitive forms also help to create the cohesiveness necessary for a large bathroom. For example, repeating the arc or curve of a tub in the design style of the vanity can tie things together architecturally.

Proportions are also very important—large spaces need larger items. This is true not only for the proportion of the fixtures but for the overall decorating scheme as well. Wall decoration should be an integral part of the overall plan. It should tie together the vertical spaces, not divide them up. Floor tile should be of a larger scale to balance out the weight of the fixtures, such as the tub and shower. Using a continuous element—such as continuing the floor tile into the shower—is often a consideration. Soft elegance is ultimately the most effective style for a large bathroom.

The Fixtures—Tubs, Showers, Sinks, and Toilets

Every bathroom remodeling project that I have been involved with seems to begin with choosing the fixtures. Why? Well, I guess it's because that's one area that is relatively simple, or so it seems. Regardless of where they ultimately will be placed, you know for sure you're going to have to have them! So, why not start by choosing something pretty?

Unfortunately, things are not as simple as one would think. Bathroom fixtures have become quite complex. From the space-age materials to the ultimate in high-tech accoutrements, choosing bathroom fixtures is more than a guessing game. I will attempt to give you the basics in as simple a format as I can. Ultimately, you will need to find a salesperson/dealer whom you trust. Shop around, take your time. Bathroom fixtures are expensive and not easily replaced.

THE TUB

If your existing tub is the ideal shape and size, and you are not going to be relocating it to another space, then consider reglazing. I have successfully had this done for many clients. There is a difference between spray painting and reglazing. Do not let yourself be talked into painting your tub. Unless, of course, you never plan to use it again. Paint will peel off!

Reglazing requires that the tub first be acid etched and then sanded smooth. After this, a porcelain finish will be chemically bonded onto your old tub. Usually the cost is around $250 and is guaranteed for two years. Be aware, however, that you cannot use abrasive cleansers to clean this new bonded finish. If you treat it gently, then it should serve you well. It certainly beats tearing out, disposing of, and replacing an otherwise good bathtub.

New bathtubs come in a variety of shapes and sizes. There are hundreds of catalogs available, and choosing one can be confusing. To narrow down your search, be sure you have specific measurements. The *standard* bathtub is rectangular, usually 60 by 30 by 14 inches. They are also available in wider sizes to accommodate two, or longer for tall guys like my husband. Typical cost is about $350. Designer models with more intricate styling and luxury features cost $700 to $2,000. The $2,000 price tag usually is for porcelain enamel over cast iron. Cast iron is still the most durable material. Enamel-coated cast iron is the most expensive because it resists staining and scratching.

Oval-shaped bathtubs are a popular choice these days. Many of them have a rectangular surround to make them easier to place. *Corner* tubs are wonderful for smaller bathrooms. Some of these are really attractive with unusual shapes. Kohler manufactures a diamond-shaped tub, making it ideal for fitting into a corner, peninsula, or bay.

Clawfoot tubs, with their vintage styling, are making a great comeback. As I mentioned earlier, *soaking tubs* are all the rage. They are avail-

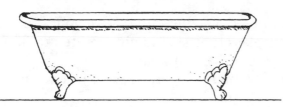

BALL AND CLAW TUB

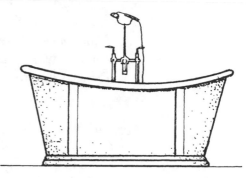

VICTORIAN SOAKING TUB

able in a variety of unusual materials such as copper, wood, stainless steel, granite, marble, and, of course, cast iron. One manufacturer, Savoy, has a tub that can be custom painted with any finish. They are elegant, artistic, sculpted, and fun in style—from sleek modern to vintage elegant. They are freestanding, set as a focal point, and demand to be the center of attention in any bathroom. There are deep enough to immerse your entire body up to your chin while resting your arms on perfectly curved rims. Some are as long as seven feet and thirty inches deep. Kohler recently introduced a cast-iron version designed for two, which is 72 by 42 inches. These new versions come with drainage systems that adapt easily to modern building codes. If you are considering buying an antique tub, it may be worth looking at new ones for this reason. Recognize that these tubs, new or old, require a *tub filler*. That's those free-standing pipes next to the tub that provide the wonderful hot water you will need to enjoy your soak. Remember to consider this little issue early on in the design process.

If you're not choosing one of the more exotic materials used for soaking tubs, then you basically have a choice of the following materials: fiberglass, acrylic, cast iron, or enameled steel, and one new material—a composite marble.

Fiberglass

This is usually the most economical and as a result the most common choice. It is plastic reinforced with fiberglass. FRP is a gel-coated lightweight fiberglass. Fiberglass is available in a wide variety of shapes and sizes because it can be molded easily. Most manufacturers offer a ten-year warranty. Fiberglass tubs should give you ten to fifteen years of good service before it shows much wear.

Maintenance. Use only mild, nonabrasive cleansers. Wipe dry, and occasionally wax to protect the finish.

Price. A five-foot-long standard tub costs around $100 to $150.

Acrylic

This is generally thought to be more durable than fiberglass. Often it is reinforced with fiberglass. It is lightweight and can easily be formed into different shapes, making it popular for whirlpools and other large tubs that would otherwise be too heavy to be practical. Acrylic is also a good insulator, which means that water will actually stay warmer longer.

Maintenance. Do not use scouring powders or any other abrasive cleaner, it will scratch. Clean only with mild nonabrasive cleansers and wipe dry with a cloth. You can occasionally wax it to better protect the finish.

Price. A standard tub in acrylic can be found for as little as $100, but I recommend spending at least $250 for a better quality.

Cast Iron

Though this is wonderful, it is very heavy! It is usually enamel coated, making it extremely durable and resistant to staining and scratching. Since it cannot be molded like acrylic or fiberglass, it is available in fewer shapes and sizes.

Maintenance. Use a nonabrasive cleaner to sustain the shine and smooth finish.

Price. Starts at $200 and keeps going.

Enameled Steel

This is the new and improved version of cast iron. It is lighter and less expensive, which gives it an advantage. However, it does chip more easily.

Maintenance. The same maintenance as cast iron; use a nonabrasive cleaner to protect the shine.

Price. Starts at $100.

Composite Marble

Manufactured by IBC (International Bath Company), it is a combination of 80 percent resin and 20 percent marble that has a gel-coat finish. The marble composite can be stretched further than standard acrylic compounds, making possible a tub that is 23 inches to 24 inches deep—a standard acrylic model is only 20 to 21 inches deep. As a result, the custom-made tub is deep enough to soak all the way up to your shoulders.

Another new idea in tubs is a flat-bottom whirlpool/acrylic tub with the classic lines of a standard cast-iron tub, made by Waterworks. This has several advantages—it is light in weight and also ideal for a tub-shower combination—unlike the standard cast-iron tub with a rounded (and too dangerous for showering) bottom.

When choosing your bathtub, it is also wise to give thought to the tub surround as well, since they will work together. Basically you can choose from tile, solid surfacing, or one-piece surrounds. *Tile* is classic and can be designed in anything from a simple basic style to an elaborate custom style. It is durable and available in any price range. The biggest problem with tile is actually the grout. (The dreaded grout!) Silicone sealers can be applied to deter stain and mildew, but it is only temporary and will need to be reapplied at regular intervals. I only recommend using a shade of gray for grout, since ultimately, this is the color it will end up. *Solid surface* materials such as those made by Corian, Avonite, Formica, Du Pont, and Wilsonart, are very popular. They are available in a large variety of colors and styles. They are durable and easy to clean. But they are not inexpensive. Be sure your contractor is an *approved* installer, otherwise your warranty will not be valid. *One-piece surrounds* are available in fiberglass or acrylic. They are relatively inexpensive and can be installed by a DIY person. They can hide a multitude of problems on old walls, which is why they are a popular choice. Their biggest drawback is that they are prone to being scratched.

Here are some tips for buying a bathtub:

❀ Just like buying a sofa or a mattress, be sure to try it out! Climb in and make sure that it fits. Do you have enough room to stretch out? Is it comfortable for reclining and relaxing? Are the sides at a height that is comfortable for resting your arms?

❀ If you like to "soak" be sure the tub isn't so big that the water is cold before you've had a chance to relax. Also check the height of the overflow drain—this determines how high you can fill the tub. If you would like a "soaking-tub-for-two," then oval, hourglass, or round shapes with center-mounted faucets are best.

❀ If the tub will be used for bathing children, consider one with higher sides, to keep the splashing inside. Some people prefer raising the tub up on a platform to make it easier on their backs.

❀ If you choose to have a skid-resistant bottom surface, recognize that they can be tough to keep clean.

WHIRLPOOLS

I have already expressed my opinion earlier about the practicality of a whirlpool tub. However, you do not have to agree with me. So, here is the basic nitty-gritty on whirlpools.

According to my wise mother, you still get what you pay for. If you decide to cut corners and price ($500 range), then you will probably end up with a glamorous version of a basic bathtub. The power of the jets will be nearly nonexistent and will do virtually nothing to ease your aches and pains.

If you are purchasing a whirlpool tub specifically for therapeutic purposes, then buy the best you can. My mother recommends two specific brands, Pearl and Jacuzzi. They have a starting price of $1,000 and can go as high as $5,000 for the top-of-the-line luxury model.

The average number of *jets* for a whirlpool is four, but the larger the tub, the more jets you need. It is not unusual to have as many as ten. It is preferable to have *adjustable nozzles* on the jets so you can direct the water flow. Higher-end systems let you choose a pulsating or steady flow and allow you to adjust the volume. The direction of the flow and the ratio of air to water determines how vigorous the massage will be. More air means a more vigorous massage. With a customized system, you can choose a stronger setting for your back and a gentler one for the more delicate areas. Kohler even has a whirlpool model

Two details that should not be neglected in designing a custom shower unit are: a seat/bench, and a ledge for shaving legs.

with a special pillow—inside it are pulsating neck jets. This is one option that really appeals to me. *Pumps* come in a variety of sizes, ranging from half horsepower to two horsepower with varied speeds. Be sure to opt for an *air switch* or *control panel* on the tub rather than on the wall. It will allow you to turn the whirlpool on and off from within the tub. Most tubs are equipped with an automatic shutoff, which will shut the tub down if the water level drops too low, thereby preventing the motor from burning out.

A whirlpool tub can hold as much as 150 gallons of water. This can be an issue in drought-prone areas. But there are whirlpools that hold as few as fifty gallons. They would be a wiser choice for such areas. Be sure you are aware of the combined weight of the tub plus water and plan accordingly for proper structural support. It is not uncommon for there to be surface cracking around the surrounding tub area after a few months of use. I have seen an adjacent shower wall crack and cause leaking in the level below as a result of the weight of one of these tubs.

Obviously, all this water will require a strong *heater* to keep it warm. Remember to add the cost of electricity to your overall budget.

THE SHOWER

It is interesting to watch the juxtaposition of showers and baths. As the bathtub reverts to the simplistic style of a soaking tub, the shower is becoming the virtual high-tech wonder—from the outrageous self-contained $15,000 model from Jacuzzi, to showers totally custom-built to fulfill all your fantasies.

The basic construction and materials for a custom-built shower unit are the same as those for the bathtub surround. In my opinion, the most versatile and easy-to-maintain materials are solid-surface products. They eliminate the need for the difficult grout

Heaters are one of the more expensive items to use. An evening of bathing can cost $5 to $20 just in electricity.

problem, mildew, while giving you the flexibility to custom-design and fit a shower to your needs.

When I built my current home, I paid particular attention to the leg-shaving ledge. Unfortunately, the *men* measuring for the glass shower door and panels did not. They just assumed the ledge should be *outside* the shower enclosure. They were not happy with me when I made them remove and replace the enclosure to accommodate my ledge inside the shower! I guess it's all a matter of perspective. They thought the ledge was for the adjacent bathtub (even though it had an additional surrounding ledge of its own) not the shower. Oh well . . . it worked out okay.

As I said earlier, I also added three corner-mounted tray shelves in one corner of my shower. They are mounted at regular intervals starting at approximately sixteen inches from the top of the shower.

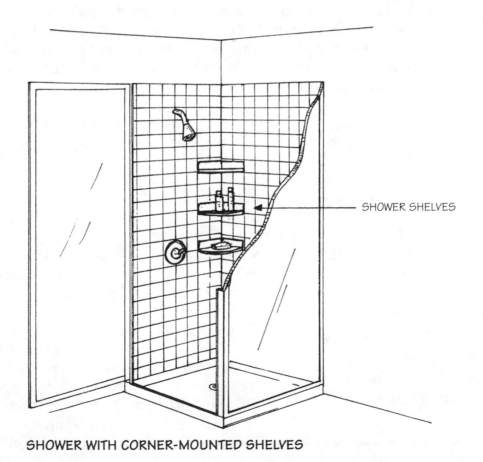

SHOWER WITH CORNER-MOUNTED SHELVES

Again, since the tile installer had never seen this done before, it took some convincing for him to believe it made sense—until it was finished! Then he thought it was a wonderful idea. It's perfect for all the bottles of "stuff" we keep in the shower—shampoo, conditioner, face cleanser, body cleanser, loofa sponge, razor, etc.

Shower Enclosures and Doors

Shower enclosures are available in many sizes and shapes. Some are designed for smaller spaces, such as the neo-angle corner enclosure. Just like a bathtub, get inside and see how it fits. Standard shower enclosures are only 32 to 36 inches square. If at all possible, install a larger one. I prefer having a shower large enough to make it possible to dry off before stepping out.

Each shower enclosure includes an array of features and fixtures from the necessary to the indulgent. So do your best not to be distracted from the basics of shape and size for your particular bathroom. The more realistic models will range in price from $400 to $1,200. I must admit, there is one luxury option by Kohler that intrigues me—a soothing foot whirlpool in the base of the shower—that sounds almost irresistible.

Shower doors have really gone through a revolution in recent years. Not long ago, there were only a few basic styles available. Now, the choices are nearly limitless. One such unit, Kohler brand Arica style, even has a round shower enclosure, which allows for maximum interior space while using a minimum of physical space. It has nonderailing door panes that slide to the sides, creating a wide center opening. A release mechanism on the bottom guide allows the door panels to swing inward for cleaning.

Another Kohler product is a pivoting door style. It has a frameless $1/4$-inch glass door that can be installed to open at left or right so that you may accommodate them without interfering with other fixtures in the bathroom. Some products also include what I consider to be an ingenious idea, compression-fit wall jambs, which eliminate the need for drilling into the tile or shower surround. Two Kohler styles, Arica and Forum, both have this unique feature available. You simply turn the adjustment screws—the telescoping jambs tighten against the shower space for an even and secure fit, even if your walls are not plumb!

Bypassing doors are one way to save space with a shower door. The newer models have detachable and removable bottom guides, which enable the glass panels to swing away from the frame for cleaning both the glass and the rail.

As I mentioned earlier, another consideration for remodeling should be having a barrier-free design, hence, barrier-free shower doors. The Helios model of doors by Kohler have a trackless style design. You simply slide the door, fold it, and then step in. They also manufacture a model for bathtub/showers to provide unobstructed access. They create a very efficient and effective way to make a bathtub/shower more accessible.

Grab bars should also be provided whenever you are addressing barrier-free design. The good news is that there are now "designer" finishes and styles available for grab bars. This makes the whole idea of safety easier to accept. Kohler also has a totally self-contained bathtub/shower unit designed specifically for barrier-free accessible use. Their Freewill model has integrated grab bars and a seat within the unit. It also features a soothing hydro-massage with six adjustable jets. This unit completely complies with the ADA.

Glass selection for shower walls and/or doors is a matter of preference. You may choose from crystal clear, mottled, decorated, or a rice paper effect. I love the look of crystal clear glass, but only in "model" homes. I cringe at the thought of keeping the crystal clear glass clear! But for those of you who are willing to spend the time keeping it clean, it's a great look. I am definitely more practical and recommend something with a textured/mottled finish.

Shower Options

If you are considering a *steam* unit, be sure to check with the manufacturer and someone with a lot of experience. Too often, "steam dreams" turn out to be expensive mistakes that never quite meet the expectations of the user. You will also need to have a *domed* ceiling and you should not choose a fiberglass enclosure. A steam unit can also do damage to the finish of solid-surface materials and acrylic. Additionally, you will need a tight-sealing door. All these factors can make this a very expensive proposition.

Saunas are another glamorous feature you may have thought about. If so, then plan for a unit independent of your shower. I have several clients who have installed such units in their basements. You will need a minimum of 4 by 4 feet of space. There are precut and ready-to-assemble units, or you may have a custom unit fabricated. A sauna brings relief for soothing sore muscles by the use of a heater with rocks on top. Water is then ladled onto the hot rocks producing humidity (not steam). This dry heat will average a temperature of about 170 degrees Fahrenheit.

Showerheads

When all is said and done, it is the showerhead itself that makes the most difference. And this is clearly an area of personal choice. Some of us prefer a strong pulsating shower, while others want the soft mist of a rain forest. All this diversity adds up—in dollars that is. They range in price from $50 for a massaging showerhead to $1,000 for a "shower tower" with adjustable sprays.

The most common showerhead is the *fixed* or wall-mounted showerhead. One of the newer innovations in fixed styles is the ceiling-mounted drenching spray. My husband informed me that they just installed these wonders in the men's locker room at the gym. (I don't recall seeing them in the women's locker room—hmmm?) Some have waterfall spouts, which create a softer cascade of water; others offer pulsating sprays at variable speeds.

I recommend installing a fixed-style showerhead to meet the necessary requirements of the tallest person in the house. For example, my husband is nearly six foot four (I am five foot two), so I had the showerhead installed at a height that would allow him to wash his hair without having to stoop to rinse it. The adjustable swivel head allows me to direct the shower at an angle that is comfortable for me. I also raised the height of the shower doors and the tile wall around the shower to accommodate my husband's height. This eliminated the problem of a wet floor or wet wall.

The most important consideration when installing a fixed-style showerhead is the height at which you install it.

Handheld showerheads are probably the next most popular style. One reason is that they are convenient for showering children (and pets). But more important, they are wonderful for the elderly. It makes it possible for someone to sit on a seat in the shower and still be able to comfortably shower

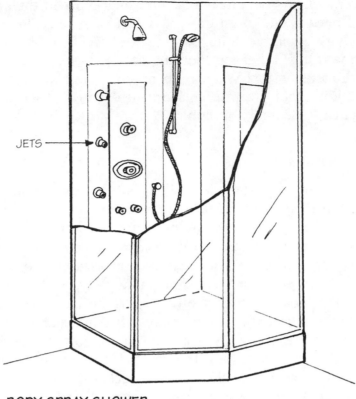

JETS

BODY SPRAY SHOWER

themselves. Some are available on a pole-mounted unit, which allows you to slide the showerhead up and down to meet your specific needs. This eliminates the need to physically hold the showerhead. (A side note for pet owners: I used to have an eighty-pound dog. The walk-in shower with the handheld showerhead was the perfect place and way for bathing her while keeping the rest of the bathroom intact!)

Body spray units provide several showerheads/nozzles mounted at various heights that spray at you from all sides and angles at the same time. The overall effect is similar to having several people shooting at you with waterhoses! If that conjures up joyful memories from your childhood, you will probably enjoy a body spray. For me, I have to admit it is not my "cup of tea." I prefer being able to seek occasional relief from the spray, which is virtually impossible with a body spray system. Imagine accidentally turning that thing on when the water is still *cold—brrrr!*

Rainbars are a more recent development. Just the name conjures up the soothing comfort of a rain forest. Basically, this is the old outdoor sprinkler let loose in the shower. It is a long metal bar with numerous little openings that give a gentle spray. They can be mounted on the wall or the ceiling of the shower. If wall mounted and an additional feature to your standard showerhead, they allow you to shower without getting your hair wet.

Ventilation

We cannot have a discussion about showers without discussing ventilation. In my opinion, ventilation systems are most effective for the removal of *odors* but much less efficient at removing *moisture*. I have yet to see a system that convinces me otherwise. That said, an effort to control moisture is essential to preventing mildew. The easiest way to accomplish this is to crack open a window while showering. Additionally, you should try to have a ducted ventilation system with access to the outside. It is virtually worthless to just install a recirculating fan. The efficiency of a fan can be improved if it has the power to actually change the air in the room a minimum of eight times each hour, or a fan that moves at least 50 cubic feet per minute (cfm) of air. To determine the proper cfm's for your space, multiply the length by the width; the total will match the number of cfm's your fan should have. Be aware, however, that some fans will generate a noise level that can be deafening, which means you probably won't use it because it's annoying. Be sure to keep the noise level down to .3 sones of sound for the quietest fan. In a larger bathroom, it is best to have a separate vent for each individual area, i.e., shower, toilet, and bathtub.

The bottom line: consult with your electrician and your HVAC (heating and air-conditioning) contractors for their suggestions. They will have access to resources that may not be available to the general retail public. They also will have insights into how best to duct their way through your home, to provide the kind of air access that is necessary to a successful ventilation system, without creating a cold air leak.

SINKS

Bathroom and kitchen sinks have a lot in common. They are basically available in the same materials with one exception: bathroom

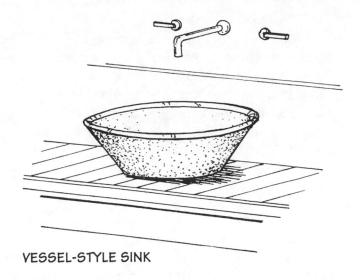

VESSEL-STYLE SINK

sinks are now available in vitreous china. It is one of the most prevalent materials chosen for bathroom sinks. It is a glossy, nonporous, glasslike surface that is very durable as well as beautiful. The biggest difference between kitchen and bathroom sinks is obviously style.

One of the newest trends in bathroom sink design is the vessel-style or washstand look. This is the use of bowllike sinks mounted on top of the counter, giving the appearance of a freestanding bowl. (Notice again the trend toward the old Victorian era.) These sinks are virtual works of art that can set the tone for the entire bath. They are available in stainless steel, faux stone, colored or clear glass, and ceramic. As a result, the choices for faucets and fittings have multiplied and changed to accommodate this trend. The resurgence of the wall-mounted gooseneck faucet and decorative handles completes the unique look.

The basic style choices for bathroom sinks are: pedestal, above counter (vessel style), console, undermount, self-rimmed, overhanging, and integral.

Pedestal

These sinks are very popular, particularly for powder rooms, where a minimal amount of storage is necessary. They are available in a variety of sizes, designs, and prices. They come in two types of material choices: cast iron with porcelain enamel, and steel with porcelain enamel. Cast iron is heavier and more expensive than steel.

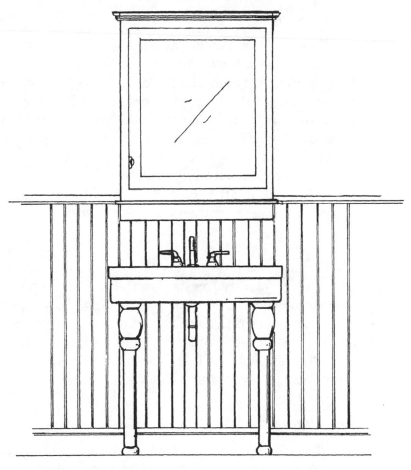

PEDESTAL SINK

The first consideration when choosing a pedestal sink should always be proportion. Most powder rooms have a niche or nook space for the sink. I prefer a minimum of six to eight inches clearance on either side between the wall and the pedestal sink. Many designers will insist on twelve inches of clearance. However, I have not found it necessary. Instead by using a larger scale proportion it creates the illusion of a larger space without hindering the efficiency or practicality of the room. Height is also important—take into consideration the height of your ceiling. If your ceiling is higher than 8 feet, then choose a pedestal that is taller. The standard range of height for pedestal sinks is 32 to 35 inches. Prices vary from one extreme to the next: $150 to over $1,000. It all depends on whether you want practicality or *art*.

Vessel Style

These sinks can be wonderful for both small and larger bathrooms. In a small bath, they can bring visual interest to an otherwise boring vanity cabinet. In a larger bathroom, they can lend artistic lines to a too-long vanity. My one word of caution: They take a little getting used to—especially if you are in a hurry. You have to remember to lift everything up above the edge of the vessel. You may, for example, find yourself knocking your perfume or makeup bottle up against the side of it. In addition, cleaning underneath the vessel will require a little extra time. All in all, though, they are beautiful and unique and can be the perfect choice for the look of opulence.

Console

These sinks are freestanding sinks that do not require a cabinet or vanity. They are supported by elegantly shaped porcelain legs. These Victorian era beauties are best suited for larger bathrooms. Their size and lack of storage would make them impractical in a smaller room. Choosing one with sufficient deck room is key. Also consider adding a separate shelf above the sink for your personal necessities.

Undermount

These sinks are attached to the countertop from beneath. *Self-rimming* or drop-in are just the opposite. They drop into a hole in the countertop, leaving an edge to overlap onto the countertop.

Overhanging

These sinks protrude beyond the counter edge. These are often wall mounted. They are specifically designed for use with a wheelchair. However, they are also ideal for someone who has a difficult time standing because they easily provide for the use of any chair.

Integral

This simply means that the sink and countertop are one smooth continuous piece. They are probably the easiest to clean because there are no nooks and crannies where the dirt can collect.

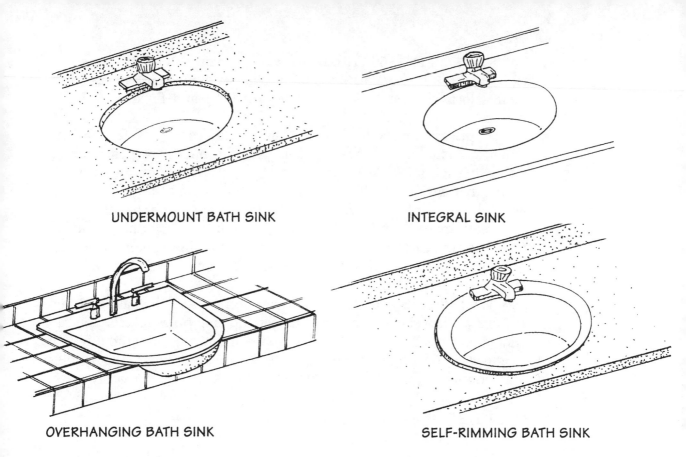

UNDERMOUNT BATH SINK

INTEGRAL SINK

OVERHANGING BATH SINK

SELF-RIMMING BATH SINK

FAUCETS

No sink is complete without faucets. (Surprise, they are a separate item.) Believe it or not, they can cost as much as the sink or even more. One of the biggest factors in determining prices for faucets is the part you cannot see—the inside. The best faucets have brass bodies as opposed to steel. Additionally, basic mechanical configurations vary from *cartridge* style, which houses parts in a metal sheath, to *compression* style, which uses a screw to keep things together, to the most recent development—ceramic disc valving. Obviously, cartridge and ceramic disc valving are more expensive (See kitchen sinks, page 127, for more details on ceramic disc valves). Of course, style will also have a huge impact on the cost. The prettier and more detailed, the more expensive. Faucets range in price from as little as $50 to $500 and up.

The $50 to $150 range will give you basic serviceability with a less expensive zinc body. Usually your only choice of finish in this price range is chrome.

The $150 to $350 price range will give you solid brass construction with the higher quality ceramic disc valving. You will also find your selection of style (one- or two-handled) and finish greatly increased.

At prices above $350 you can begin to choose not just for function but for artistic form as well. If you have big bucks to spend, take a stroll through a Sherle Wagner showroom. Here the entire bath is a total experience. Their line of sinks and faucets are fit for a king. They are truly works of art—investment quality art—and offer handles made of crystal, marble, and filigreed gold. They are not the only manufacturer offering this array of choices; many of the more common household brands are adding luxurious models to their lines. After all, this is the "era of the bath."

While dreaming of all that glitters, don't forget the basics. When replacing a faucet, make sure the faucet and sink holes match. You can choose from *center set,* in which the spout and handles are mounted on a central connecting plate, or *single-hole* fixture style, where the handles are mounted directly on the stem of the faucet, or *spread-set* style, where the handles and spout are set into separate holes in the sink or counter. The distance between sink handles is either 4 or 8 inches.

THE TOILET—THE MOST IMPORTANT SEAT IN THE HOUSE!

If you haven't purchased a new toilet in some time, you may be surprised at a new congressional law that affects toilet operations. (No, I'm not kidding!) You may use only 1.6 gallons of water per flush. (The old-fashioned toilets used up to six gallons of water per flush.) And up until very recently, this was creating some not-too-clear problems. The plumbing industry struggled to build a toilet that could efficiently work with only 1.6 gallons of water. Unfortunately, I was a victim of this situation. I built my home about the time the new law took effect. As a result, the toilet in my powder room is not very efficient. Although it uses only 1.6 gallons of water per flush, I need to flush two or three times to actually clean the bowl! How efficient is that?!

The good news is that if you are reading this section, you probably have not yet purchased your new toilet, and the newer models are very good at working with only 1.6 gallons of water. The more recent topic

Another standard being used to determine the worth of a toilet is the toilet paper flush test. Kohler's Wellworth *model boasts being able to handle the equivalent of a 280-sheet roll of toilet paper in a single flush. Now that's an impressive performance!*

of discussion has been trap size. For those of you not familiar with plumbing lingo, trap size refers to the diameter of the object that can easily pass through the system. Kohler, for example, discusses this process by comparing the ability of their Wellworth trapway being able to handle an object the size of a handball, while the competition can only pass a golf ball. As their ad states: this is 26 percent larger than ANSI (American National Standard Institute) minimum standards!

There are two basic flush methods available: gravity fed and pressure assisted. *Gravity fed* is the traditional method where water is stored up in the tank and rushes down, forcing everything out through a built-up pipe. This pipe curves up and then down. It fills with water, which keeps gas from entering the house. A standard two-piece, gravity-fed toilet in white or almond starts at about $160. *Pressure-assisted* units have a tank inside the tank. This holds trapped, compressed air. The air acts like a spring to shoot the water forward. This is the fastest flush around—emptying the bowl in less than four seconds! Pressure-assisted systems price in the range of $200 to $500. An elongated bowl will raise the price. A one-piece pressure-assisted unit can cost $600 and up. There is a *third* option available—using an electric pump. This creates pressure quietly. The average price of a pump-assisted unit is $900. Kohler has introduced a unique idea that may catch on with others. It is a dual-flush system. One lever uses 1.1 gallons of water for liquid waste, and another lever, for solid waste, uses 1.6 gallons of water per flush. Price: $170. Kohler's most recent introduction has a *heated* seat! That's right, you no longer have to suffer the shock of a cold seat in the morning! Their new French Curve model fits most elongated bowls. It comes in twenty different colors. Prices begin under $100.

Toilet designs are available in many styles: freestanding, one piece, wall hung, corner, and low profile. *Freestanding* models have a tank sitting on top of the bowl. The bowl is mounted on the bathroom floor. The old-fashioned version of this style had a wall-mounted tank. *One-piece* models are similar to freestanding except that the tank and bowl are a single piece, making them easier to install. *Wall-hung* models require a strong support in the wall. The soil discharge pipe goes through the back wall instead of the floor. This style is popular in health-care facilities because it is easier to clean around. *Corner-style* toilets are perfect for tiny bathrooms because their tank is triangular and fits neatly into the corner. But they can be difficult to find. The *low-profile* model is the newest and most dramatic style. The tank is only slightly higher than the bowl.

A new introduction into the residential bathroom is the higher profile toilet. It has a bowl height of $16\frac{1}{2}$ inches. This additional height requires less bending, so it is more comfortable for people who are taller or who may have limited movement. Kohler's Highline model complies with ADA standards.

Prices for toilets run the gamut—from under $200 to as much as $1,500 for an artist-designed suite. Yes, even toilets are now available in hand-painted patterns to match your bath. High-fashion colors are usually more expensive than the basic neutrals. In my opinion, it makes more sense to stick with the basic neutrals anyway. They won't "date" your house. It's amazing how easy it is to tell when a home was built or remodeled by the color scheme of the bathroom fixtures. For example: pink—1950s–60s; rust—1970s; black or mauve—1980s; gray and taupe—1990s.

BIDETS

A bidet resembles a toilet, but it has hot and cold running water for bathing your private parts. For some strange reason, American men seem to laugh every time the word *bidet* is mentioned. I'm not sure why, nor do I think I really want to know. Nonetheless, I think it's unfortunate that we Americans have been so slow in accepting them as an important feature in the bathroom. I believe that every woman should have one. Unfortunately, our American bathrooms are only beginning to be designed to accommodate this luxury.

There are many things that I find Europeans do better. They understand the need for a balance between work and pleasure. Their holidays (our version of vacation) are extended and more frequent. Their lunch hour lasts longer than an hour. (They make up the difference by staying later.) Their cities are designed with people in mind. It's easy to find an inviting outdoor place to sit and become engaged in conversation. Many of the old cities are open only to pedestrian traffic, making people a priority instead of cars. And, of course, their bathrooms have *bidets!*

This European approach to living is gradually catching on in the States. Although bidets are still considered a specialty or luxury item, they are beginning to appear in the regular product lines of bathroom fixtures. Bidets are usually chosen to match the toilet. They are installed, in most cases, next to the toilet. A towel bar and soap dish

BATH CABINET WITH MIRROR, LIGHTS AND WINDOW ABOVE

should be located nearby for convenience. Bidets are priced a bit higher than toilets, starting at $200.

Some manufacturers, such as Toto, have developed dual-purpose units to capture the bidet market without the need for a second fixture. Their Zoe model toilet features a toilet seat that has heaters and a warm water spray, as well as an aromatherapy spray.

On that fresh scent, remember the whole point of this chapter—to create a bathroom that will help you put your best-(scented) self forward. Renovating or building a new bathroom is exciting and intimidating at the same time. It's important to take the time to plan and choose all aspects carefully in order to make it aesthetically, functionally, and financially worthwhile. If you feel you are in over your head—ask for help. You can save time, money, and hassle by working with an experienced professional.

Words of Wisdom

❀ A telephone in the bathroom is becoming a standard fixture. If you decide not to go with this new trend, then at least consider an intercom system.

❀ Do not use a striped wallpaper pattern in a small bath—it will only make it feel smaller!

❀ My mother says (and she should know, as a kitchen and bath specialist) the most common question people ask is: Can they install a "tub kit" (complete ready-to-install replacement walls and tub cover) over existing tile if the tile is pulling away and coming off the wall? The answer is NO!!! This is because if you attach the new tub kit to the existing tile, it is only a matter of time before the new tub kit also pulls away from the wall with the tile. You MUST cut out the old wall, address any water problems, install new "green" board (waterproof drywall), and then you may install the new tub kit.

❀ The NKBA recommends pressure-balanced (antiscald) valves for your shower and sink faucets. This allows you to preset the hottest temperature the faucet can deliver, regardless of the temperature setting of your hot-water heater. This is a good idea not only for families with children but also for the elderly.

❀ NKBA is a great resource for information and products: 800-401-NKBA

❀ Another tip from Mom: If you are replacing a toilet—measure the *rough*. Rough is the distance from the wall to the closest hold-down bolt of the toilet. Normal rough is 12 inches. However, there are also 10- and 14-inch rough toilets. A 10-inch rough can actually measure between 9½ and 10 inches. A 14-inch rough can measure between 13 and 14 inches. In addition, you need to know whether your existing toilet is a one-or two-piece unit. One piece units are only available in 12-inch rough. Toilets that have a 10- or 12-inch rough are two-piece units and are more expensive than one-piece units. (A 12-inch unit starts at $120; 10- or 14-inch units start at $180.)

❀ A base price porcelain on steel bathtub costs about $100. However, I do not recommend purchasing one this inexpensive—it will not hold up and it will be very noisy because it has virtually no insulation.

❀ American Standard makes a tub called Americast. It is an insulated metal backing with porcelain. It is well priced and definitely a better choice than the cheaper models.

❀ Cast-iron bathtubs *are* still available, contrary to popular belief. The cost is similar to Americast but it weighs twice as much. If you insist on cast iron, then be sure you've got the structure to support it (when it is filled with water).

❀ A recent development in shower design is the new trackless shower door. Available from Kohler, Majestic Shower Design, and Glass Products Institute, the nonderailing door panels slide to the sides, creating a wide center opening. A release mechanism at the bottom allows the panels also to swing inward for cleaning. Some models are equipped with a full 180-degree rotation. This new system is completely compliant with the ADA (Americans with Disabilities Act) standards, making it a good choice if you are middle-aged or older.

❀ When planning a bath be sure to consider "soundproofing." I have been in very expensive homes in which you can hear in the dining room the toilet flush! Talk about ruining your appetite!

❀ The AARP (American Association of Retired Persons) has some materials on home improvement ideas for seniors. For further information call: 202-434-2277.

❀ Designing for acoustic comfort is much more than just separating high-traffic areas from quiet zones such as bedrooms. Owens Corning's Quiet Zone system, which uses acoustic batts stapled into interior walls for sound absorption, is one good choice.

❀ Faucets with lever or wrist blade handles are much easier to operate because they require a minimal amount of manual strength and dexterity.

❀ Elegant new handles can be an easy and inexpensive way to transform the appearance of an old bathroom vanity.

❀ Try using a single-panel, 80-inch-high-by-20-inch-wide shutter at the bathtub to hide the shower curtain and showerhead. Add a shelf above, and it's the perfect camouflage for the shower curtain rod.

❀ Make sure your bathtub and shower are easy to get in and out of. I do not recommend steps beside a whirlpool tub—they are too much of a hazard.

❀ Never attempt to directly step into a raised tub—it's just too deep and too dangerous! The proper way to enter or exit a "raised" tub is to sit on the side ledge and swing your legs (pivot).

❀ "Closeouts" are a great way to find bargains, however, be sure the products you are getting meet new standards. As of August 6, 1998, all kitchen and bath plumbing inventory must comply with the Safe Drinking Water Act. The Act requires all faucets and fittings used to dispense water for human ingestion to be *lead free*. Products affected include kitchen, bath, and bar faucets, hot and cold water dispensers, and ice makers. Those products in compliance should carry a certified mark of NSF or UL.

❀ Locate shower valves to the side of the showerhead rather than directly below so the temperature can be adjusted without reaching through water that is too hot or too cold.

❀ Here are the answers to the two most-often-asked questions: Where to hang the toilet paper holder? Answer: If possible, in front of the toilet within reach of a seated person, 26 inches from the floor. Which way should the toilet paper be placed, with the paper pulling forward or backward? Answer: (I think forward, because it makes it easier to see and find the end of the piece.) Whichever way works for you—just be consistent.

❀ Microban, an antibacterial material used mostly for commercial and health-care applications, has recently been introduced to the residential market. Microban reduces mold and mildew buildup and inhibits the growth of bacteria. It's a protectant that when added to plastic surfaces during fabrication actually becomes part of the finished product. Aqua Glass is now producing bath fixtures containing Microban. Products with Microban cost about 10 percent more than those made without it.

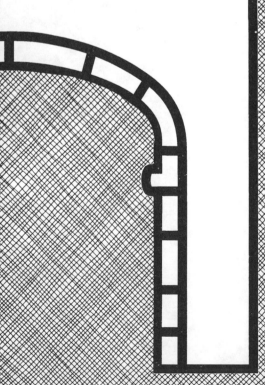

7

LIVING, DINING, AND FAMILY ROOMS

Expectations and Evaluations— Rearranging and Redesigning

Living, dining, and family rooms are the *public* spaces of your home. These are the spaces where life happens. We live, dine, entertain, relax, and reconnect here. If you could wave a magic wand and instantly change these spaces, what would they look like? What could you change about them that would not only enhance the comfort level but also the space itself? Would you raise the roof? Would you bump out a nook? Remove walls that separate two rooms? Maybe a whole new addition is needed. Then again, maybe windows and skylights are the answer.

Allowing yourself to take an imaginary trip into the "perfect" space is a great place to begin planning a project. Albert Einstein said, "Imagination is more important than knowledge." I agree with him. These little daydreams are the makings for discovering your options. For the moment, assume money is not a problem. The point to this exercise is to give yourself the freedom to truly imagine all the possibilities. It is from this reference point that you will be able to specifically discover the true problems and faults of these spaces while discovering a variety of solutions, desires, and creations to improve the quality of your home and your life. After all, isn't that the point? Your home should be your haven, not just a museum. Recognize that ultimately your home should be designed around you and your family's lifestyle, not the other way around. A recent survey found that most Americans want quality and value, not glitz. We want a place to live in comfort-

Your home is meant to be lived in. It should complement your life, not complicate it.

ably, not just a place to show off. To truly improve your home as an investment as well as an amenity, you should complement or improve on its original style. The workmanship should equal that of the original house. (Unless, of course, you are improving the overall quality of the entire home.) Since we move less often than over the past two decades, we want more flexible floor plans. Such plans can be adapted as easily as our lives and priorities change. The space should function well through late-life stages (a bedroom on the first floor) as well as now, when you may have children. Many homeowners are opting for home offices where both adults and children can work. Parents want to supervise homework and Internet surfing.

Architectural Elements

Strive for accomplishing the best for your particular situation. Sometimes a *minor* miracle is all that is necessary. Little changes can make huge differences. A bright new up-to-date color paint job may be all that's necessary. Then again, maybe a combination of new flooring, cabinetry, railings, and windows may be enough to bring your home from the 1970s to the twenty-first century.

Here are some specific thoughts to ponder while you dream:

It is important to recognize that sometimes "perfect" isn't possible. Perfect may, for example, make your home too expensive for the rest of the neighborhood. Or perfect may not fit within the building envelope or building code.

❀ Making a home seem bigger doesn't always mean adding an expensive addition. In fact, you can accomplish the look and feel of spaciousness by opening spaces from one room to the next. One of the more popular areas of the home where this makes sense is between the kitchen and family room. If you don't have a family room, consider opening the wall between the kitchen and living room, or even the kitchen and dining room. For one client, we rearranged rooms. By moving the dining room to where the living room was, and moving the living room to the old dining room, we were able to create a wonderful great room/ kitchen suite. With as little time as we have together with our families it makes sense to create as much opportunity for togetherness as possible. An open kitchen/great room or kitchen/family room is a perfect way to blend the best of both. After all, isn't the kitchen where most people prefer to be anyway? Why not make it comfortable?

❀ Use the space you have. Rethink your dining room. Instead of using it only a few times a year for special occasions, why not transform it into the ultimate place for working and living? It can easily double as a home office/homework room, library, eating area, and storage room with built-ins.

❀ You can create a bigger-looking space by installing new glass doors where only windows or a wall stood before. Add transoms, or windows above patio doors and you've added even more illusional space. In addition, the added daylight will give the room a new sunny disposition that will enhance every aspect of the space, including your own disposition.

❀ One of the most annoying and confining problems with older homes is traffic control. By defining and correcting serious traffic flow within a home you can go a long way in establishing spaces whose form is designed to function.

❀ Hallways are often a waste of space. Instead, let them be a source for additional square footage—open them up to enlarge the real living space. If necessary, you can still define traffic patterns with the use of columns or half walls while still adding volume and visual space to the rest of the room. I used this technique at the end of a dining room adjacent to a passageway for a client. We removed the non-load-bearing wall. Then we added one half wall and two columns. The result—a larger-looking dining room that also has the advantage of additional light gained from the windows in the room across from the old hallway.

❀ Use your nooks and crannies. It is amazing how much space is used just as a pass-through. By adding bookcases to a wide passageway, you can realize the dream

One of my favorite ways to give the illusion of space is to play with odd angles. The old adage of "thinking out of the box" is true for homes as well. Often, the use of angles in the shape of an addition, or even a kitchen island, will allow for the designing of larger spaces and/ or additions because they are more efficient at using space.

of a library. Or create that special place for your favorite craft or hobby.

❀ Another favorite space gainer, which is also financially easier to handle, is to build a *bump-out*. A bump-out is really a "bay"—of sorts. It can be any shape—square, round, angled, and of course a standard walk-out bay. By adding just these few square feet to a window, you can completely change the appearance of a space.

❀ If you have a one-story or a split-level home, consider raising the roof. By raising the roof you can gain a two-level living space. Add a step-up and you've created a dazzling new dining or family room. Or you may prefer a loft-haven for your home office or a new guest room. This too is possible by raising the roof. Raising the roof can be easier and less expensive than an addition while still accomplishing the same goal of creating additional living space and allowing more light. The additional wall height can be filled with windows to create an unending source of light and open the room to nature.

❀ Consider creating a vaulted ceiling with additional windows or skylights. The volume of space created by lifting the ceiling will give you the feeling of larger, brighter spaces.

❀ Do you have a room that's too large? Many older homes were noted for their cavernous living rooms. With today's lifestyle, we hardly even use the living room. My sister is faced with this situation. Though her living room is enormous, her family room is just adequate. Redefining, relocating, and rethinking these spaces can make your home work better for you and your lifestyle. In my sister's case, the kitchen could be improved and enlarged by taking some of the excess space from the living room. The remaining space would be perfect for a music and sitting room.

❀ Another option for a room that is too long for its own good is to use interior glass doors to subdivide the space. This can be the perfect solution. When you need a larger space you simply open the doors. In addition, you won't lose a lot of light by closing off windows that bring light into the room. You can use traditional French glass doors. For a more contemporary look, try using pivoting wood-framed panels with diffused glass inserts. Choose from etched, sand-blasted, stained, or leaded glass.

❀ An addition is often a real consideration. When most people think about an addition, they think of going toward the back of the home. Sometimes, it makes more sense to do a *side* addition. A modest

BEFORE ADDING SECOND FLOOR

AFTER ADDING SECOND FLOOR

Living, Dining, and Family Rooms • 197

FRENCH DOORS USED TO DIVIDE UP LONG ROOM

15-by-22 space can make a home seem twice as large, particularly if you have a narrow home. The ideal space is square. So the squarer you can make your spaces feel, the better.

Consider a 1½-story addition rather than a 2-story one. This can make it easier to adapt it to the existing house.

❀ Redundant spaces can also be a problem in older homes. Redefining such spaces as porches and patios can add much-needed year-round living space. Even a three-season room such as an enclosed porch can make a difference.

❀ Built-in shelving and cabinetry and a new mantel and railings can completely change the *character* of a home from one style to another. Having realistic expectations for this transformation is critical to its success. For example, a basic cottage-style home can be transformed into an elegant Victorian. A 1950s contemporary ranch *cannot*. But it could become an up-to-date sophisticated beauty.

❀ Add a fireplace. This can be a real bonus. With the availability of ventless gas fireplaces, you can put a fireplace virtually anywhere. These advanced fireplaces are highly efficient and have no need for a chimney or electrical connection. As a result, they can be installed in rooms where no fireplace now exists. They can also be used to convert an existing wood-burning fireplace into a more efficient zone heater. There are even see-through peninsula models that are perfect between a kitchen and family room. Nothing warms up the feeling of a room better. The ambiance of a fire can transcend all the storms of the day. It's romantic and inspiring.

❀ Woodstoves are particularly popular in colder climates. The reason is that woodstoves burn slower and more evenly, therefore less heat escapes up the chimney. This makes them more efficient than a wood-burning fireplace. And unlike most vent-less gas fireplaces, a woodstove can be a real source of home heating. They can be vented through an existing flue, or through their own freestanding chimney

ADD A FIREPLACE FOR AMBIANCE

through the roof or a side wall. Woodstoves are rated by the EPA (Environmental Protection Agency) for emissions. Check out how your choice placed in rating to be sure it meets your own standards. Be aware that both the outside of the stove and the flue get extremely hot. This means that you should not plan to place furniture too close. If you have young children, put the woodstove in a room where they do not necessarily spend time alone.

❀ Moldings, trim, and details—these are three of the easiest and least expensive ways to add architectural interest as well as the look and feel of quality. New technology makes it possible to have incredibly detailed, hand-carved-looking moldings for much less cost than you can imagine. It's because many of these moldings are in fact foam, not wood. Therefore, they are manufactured, not carved. Recently, I was doing an apartment makeover for a TV show. I used a product specifically developed for commercial applications. It's actually Styrofoam, which you first coat with a textured sealer, then paint. The result is the look of cast stone. Another option would be to use several *stock*-style wood moldings and combine them to create unique custom-looking pieces.

❀ Interior doors can add distinction and quality appeal to your home. Doors with interesting detail and style can add distinguishing character. The most popular interior door is a solid-core door with recessed panels. A recent trend is the use of *old* doors in *newer* homes. It's a great way to add warmth and character as well as uniqueness to your home. Hardware (door handles and hinges) can also add a great deal of beauty and interest without adding a lot of cost.

❀ Of course, don't forget the wonders of *paint*. One of the most effective and economical ways to improve your home is to paint it. If you have a room that's dark and dingy with outdated paneling—paint it! It's amazing what a difference a coat of paint can make on paneling. By painting the paneling, you get the best of all worlds—the warmth of the wood's texture and the crisp, bright color of paint. There's even a new paint that's energy efficient. *Radiance* paint is designed to keep your home more comfortable—cooler in summer and warmer in winter. It was originally designed to keep radiant heat inside army tanks, which prevented heat-seeking missiles from finding them. It is now available for home interiors. For more information call 800-766-6776

❀ Color can also make you happy—in rooms that have limited light, the warmth of color can actually be more important than whether

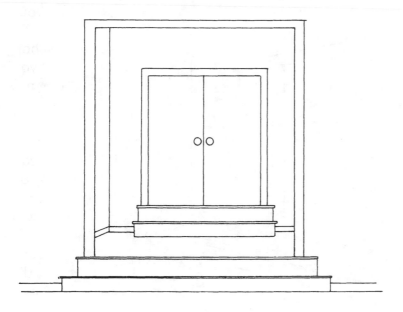

BEFORE AND AFTER ELEVATION (ADDING MOLDINGS)

BEFORE AND AFTER (ADDING MOLDINGS AND BOOKCASES)

the room is light or dark. Choose colors that make you smile. Today's homes with their open plans can make it difficult to determine where to switch from one color paint to another. To solve this problem, Leslie Harrington, of Benjamin Moore Paint & Company, suggests using a variety of graphic patterns such as stripes, squares, and free-form curves to create a series of interesting two-color walls. The idea is to allow the major color from one section of the wall to become an accent color in the other area. Harrington does recommend choosing a logical spot for the break such as the beginning of a dining ell. One suggestion is to use a color-block pattern in which you can creatively change from one color to the next.

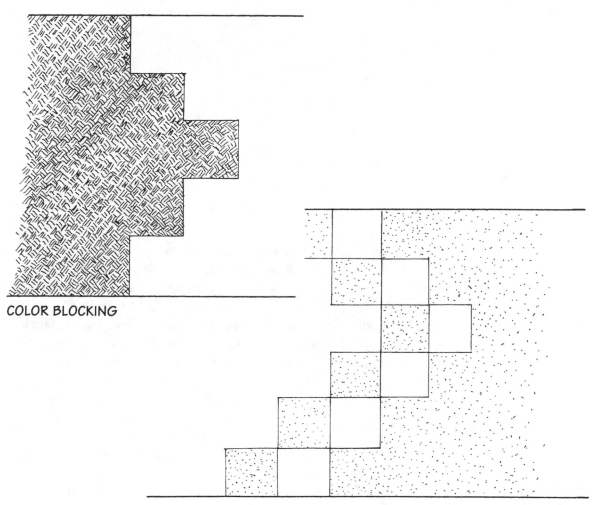

COLOR BLOCKING

COLOR BLOCKING WITH PAINT TO TRANSITION A WALL (REVERSING OUT)

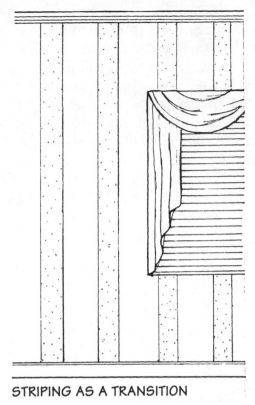

STRIPING AS A TRANSITION

The Plan

Whether you decide to work with a professional designer or DIY, plan ahead before you start planning. In other words, create a wish file. Clip pictures of houses and rooms that you like. Use my *Ink Blot Test* approach (see chapter 1 of *My Name Isn't Martha, But I Can Decorate My Home*, Pocket) to determine style, color, and formality. Using fabric swatches or wallpaper patterns, simply ask yourself if you like the swatch or not. Don't analyze it, and don't worry about where you will be using it. The point is to determine what innately is right for you. Put the swatches that you like in a pile. When you are finished you will be surprised to find similarities among your selections. You may, for example, realize that you've chosen several plaids or stripes. Perhaps you've chosen only textured styles with little or no pattern. Allow yourself to respond with your heart and your emotions.

A good floor plan is the key to any remodeling project.

Don't spend a lot of time on the samples. If they make you smile, then stick them in your wish file.

Step Two. When you're ready to start talking about planning your project, begin using the process of elimination to narrow down the number of pictures to a reasonably sized collection. At this point, you may begin analyzing them, but only from an emotional point of view. For example, is it the flooring, the windows, the color, or the furniture that most appeals to you? By focusing in on the specific area that attracts you, you can begin to choose finish materials that will make sense for you. Perhaps it's terra-cotta tile that tickles your fancy. Or maybe it's the beautiful moldings and mantel that appeal to you. By connecting your positive reaction to specific items, you will be able to ascertain the style, color, and level of formality that's right for you.

Step Three. The next step in the process is to make a list of all the things you don't like about your home. ("Everything" is not an option for an answer!) The goal to this part of the exercise is to be realistic about what can be changed and prioritizing them into what you can and cannot live with. Start with the things that absolutely must be changed. Being specific and dealing with absolutes will make it easier to determine what you really want to accomplish. Your budget will ultimately be made up of these specific items. Listing them according to your priorities will make it easier to decide what to cut out when you start to go above budget.

Step Four. Put the pencil to the paper—the blueprint. It's time to start sketching. One theory is to work in circles. Start by using circles to represent rooms. Then begin drawing smaller circles within the larger ones to represent specific tasks or work areas. This is the opportunity to play with and rearrange your existing spaces. For example, try moving the dining room to the den. Or maybe you can gain a powder room by relocating or eliminating a coat closet. Play the "why" game. By asking *why* each time you decide on an aspect of the plan, you will be forced to think more creatively. For example, *Why* do we need a wall to separate the foyer and dining room? Is there another option? Maybe the space can be sufficiently defined by just changing floor covering, thus creating a larger more voluminous space for both rooms. *Why* does the fireplace have to be in the family room? An open-hearth kitchen fireplace may be the perfect place if you spend a lot of time as a family at the kitchen table. *Why* can't we put the computer in the laundry room? After all, the laundry room is not being used most of the time. You will be amazed at the creative and much more interesting ideas that this process can evoke.

Step Five. Begin to lay out the details and the furniture. I am amazed at how many people will go through an entire building project without ever laying out a floor plan for furniture, appliances, or electronics. What's even more amazing is their surprise when they discover there isn't an electric outlet exactly where they need it, or that their antique sofa will not fit in the space where they had planned to put it! Room planning must include laying out not only the walls for rooms but also built-in cabinetry, wall systems, entertainment units, furniture, artwork, electronics, video equipment, and even plants. Take the time to decide what furniture is staying, and include it in your plan.

This is also the time to finish the lighting and electrical plans—after you have determined where the furniture is going to be placed. This way you can be sure to have the light and the electrical sources where they are needed. By taking your time at this stage of the project, you will eliminate the need for hasty decisions later that can be critical to the success of the project.

Step Six. Test-drive your plan. I always recommend that clients use props to get a better picture and feel for the changes that are proposed. Use tape to outline new dimensions and features. Move furniture around to simulate the new plan. Newspaper cutouts also work well for defining and filling the spaces to represent your new choices. I have even used the newspaper-cutout technique for clients who are purchasing new furniture and are having trouble visualizing the layout.

If you're adding on, go outside and define the new space. Stand in it, observe the location of the sun at different times of the day. What's the view like? Live with it for a while. This will help you to really get a *feel* for the new space. Be sure you are comfortable with all aspects. Keep in mind that a floor plan can feel crowded on paper. The reason—many of the objects in the plan will not actually impair or hinder your line of vision. Half walls and lower height furniture and objects are not in reality an obstruction visually, although they appear to be so on paper. It can also be difficult to *feel* windows' openness on paper. If you're having trouble with this on paper, erase the lines that indicate the window on

Don't let a contractor bully you into proceeding before you are ready. Remember, this is your house and you're in charge.

the plan. This will help your mind to see through the window, rather than hitting a wall.

If you are still having trouble visualizing the new plan, it may be necessary to have color renditions and elevations drawn. This will cost $250 to $500 but can be worth the investment if it saves you serious money later. Virtual drawings can also be helpful—CAD, computer-aided drawing. Many builders, designers, and architects are equipped with computer software programs that allow you to take a virtual trip through projected spaces. Another option is to have a scaled-down model built. I only recommend spending the money for this if you are doing *major* work. It takes a special architectural artist to put a model together. The good news is that they do make rather interesting display pieces later.

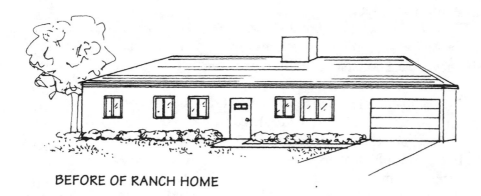

BEFORE OF RANCH HOME

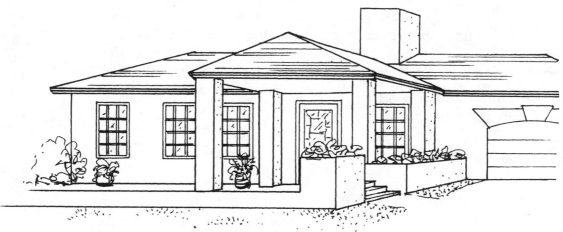

AFTER ADDING NEW PORCH, ETC.

FLOOR PLAN CHECKLIST

Before finalizing plans and estimating all the cost involved, check it out. Here is a list of things to look for:

☐ *Architecturally.* Does the plan fit the house? Does it incorporate characteristics and features from the rest of the house? Have you maintained an exterior balance by using strong details and shapes from the rest of the house? Is it proportional to the rest of the house? Most people are interested in enhancing their particular home's architecture and period. This is most easily accomplished by selecting materials that are in keeping with the original architectural style. For example, a 1960s home would benefit from a stainless steel countertop in the kitchen while a granite countertop would not be as compatible from an architectural standpoint.

☐ *Style.* Did you accomplish what you set out to do: add light, create a Victorian cottage, or a contemporary work of art? Is the mood of the new space consistent with your goals?

☐ *Lifestyle.* I'm a real stickler on this one. A perfect plan is only worthwhile if it makes sense for you and your family's life. Is there sufficient storage for children's toys, games, or your hobbies?

☐ *Function.* Have you planned your space for the way you function. I, for example, leave shoes and boots in the garage. Having a shelf to put them on makes my life easier and neater. In your attempt to add space have you created pockets of dead or wasted space? If so, how can it artistically be utilized that makes sense and is efficient?

☐ *Traffic control.* Does it make sense? Can people easily move from one space to another without creating a disturbance? Have you chosen the best wearing materials for high-traffic areas? The last thing you want is a high-maintenance high-traffic area!

☐ *Obstructions.* This is the time to check all the doors—no, not the locks, but the doors themselves. When opened, are they in the way? Doors, drawers, and even reclining chairs can be an obstruction and a hazard. Any item with moveable parts should be checked for maximum clearance.

☐ *Storage.* Add storage wherever you

Organization is the key to a simple and stress-free life—a place for everything and everything in its place. I know this is a cliché, but it does make sense.

can. Use shelving, cubbies, drawers, and cabinets. The smallest space can be used to store a multitude of stuff!

☐ *Lighting, switches, and control units.* Often the placement of electrical switches, temperature control units, alarm system control panels, air duct return and heat vents, and intercom panels are never drawn on the plan. As a result, the electrician/contractor installing them just puts them where *he* thinks they should go. This usually means exactly where you absolutely *didn't* want them to be placed. I purposely lowered many of my light switches to accommodate large artworks. Take the time to discuss and plan where the most convenient place is for all these little items.

☐ *Volume.* "Turn down the sound"—this is the time to be sure you have insulation where you need it. Consider all the elements that can give you an audio headache—clothes washers, dishwashers, video games, loud music, musical instruments, television, etc. Walls and ceilings are actually rated according to their sound-transfer performance! The key rating is called an STC (sound transmission class). This test rates the walls' or ceiling's ability to block or muffle voices and similar sounds. The higher the number, the more effective they are in buffering the sound. For residential walls, the recommended STCs are rated according to the specific room. For bedrooms, an STC of 40 is considered good, 45 to 55 is very good, and above 55 is excellent. Living rooms should be a couple of points higher. Bathrooms and kitchens should be at least five points higher. Unfortunately, most conventionally built walls have poor STC ratings, ranging from 15 to 35.

A less-used rating called IIC (impact insulation class) rates a wall's or floor's transference of impact noises such as bumps and knocks—the boys wrestling in the living room. A rating of 50 or higher is considered acceptable.

I have even suggested the "scream" test. Stand in the hallway or at the top of the stairs and shout. You will be amazed at how sound will carry, especially in rooms with high ceilings.

Simply separating bedrooms from the rest of the living space is not the only thing that can be done to *quiet* your home. New acoustic methods and products can prevent sound from reverberating throughout the house. Fluffy acoustic batts (a sheet of synthetic insulation) stapled into interior walls can absorb sound and isolate it. They are designed to fit between studs that are 14½ or 22½ inches wide and 3½ inches thick. Most have an R-11 or R-13 insulation value, the higher the R-value, the better it insulates.

Kraft-faced (brown paper covered) batts are the friendliest to handle and easiest to fasten in place. They should be installed tightly between the framing members, and snugly around such things as pipes, ducts, and electrical boxes. Leaving even the smallest portion of a wall or ceiling uninsulated defeats the whole purpose. Their effectiveness is proven with the STC test. A conventional wood-stud wall packed with insulation batts yields an STC of about 38! This number can be boosted even higher with the use of metal studs instead of wood. Using 2½-inch metal studs will give an STC yield of 45! This also works extremely well for insulating sound in high-traffic areas such as stairways. By using the acoustic batts to insulate underneath a staircase, you can ensure a quieter patter of feet on the steps. Other techniques such as staggering studs and/or using metal resilient channels can reduce contact between drywall and studs. This will prevent vibrations (especially vibrations from loud music!) from traveling room to room through the walls. You can achieve even better performance by applying a second layer of ½-inch gypsum wallboard to one side of the wall. This adds more mass, making it less prone to vibrate and transfer sound waves. Another option is fiberglass duct board instead of metal ducts. This eliminates the *broadcast* effect of sound in your home while insulating the interior.

One of the biggest openings in a wall is the doorway. Standard *hollow-core* doors are very poor at blocking sound. *Solid-core* interior doors will help contain sound within a room. Any particleboard core, composite core, or solid-wood door will work much better at providing a sound barrier than hollow-core doors. But this only solves part of the problem. Most of the sound comes from around the edges of the door. You would need to install weather stripping to really provide a seal. Think about placement of doors. By staggering doors along a hallway and arranging their swing so they don't deflect sound into adjoining rooms, you can greatly improve the sounds of silence. Another tip for silence is to avoid using sliding, bifold, and pocket doors when dealing with noisy areas.

Think about sound control when choosing finish materials for your home. Choosing *soft* ones—padded carpets rather than hardwood, tile, or laminate flooring—will minimize sound from bouncing around.

☐ *Maintenance.* This is something that needs to be considered throughout the entire process. With each selection of finish materials, such as flooring, countertops, carpeting, wall covering, and even the color of things, you must consider how easy or difficult it will be to maintain. I suggest obtaining the manufacturer's recommendations for cleaning and maintenance. Be aware that most manufacturers will not honor warranties if you have not followed and documented their recommendations. Carpeting is one such item, particularly if it has a soil-repellent applied. The manufacturers are very specific about how and how often to clean the carpet in order for it to still be under warranty. If no such instructions exist, then seek out the opinion of at least two professionals who have experience working with that specific product. Marble is a good example of a difficult product to maintain. It requires polishing, which in turn makes it slippery and therefore dangerous. Unfortunately, many "tile" stores are not fully aware of marble's necessary maintenance requirements. It is important to speak with a "marble" specialist to be sure you know what you are getting yourself into.

☐ *Safety and building codes.* Does your new space meet all the building codes set by your community? Does it meet with your own personal codes of safety? Too often, a building code issue can be overlooked until the project is complete. By taking the time to check and double-check, you could save big bucks and a lot of aggravation later.

Budget Balancing

Balancing a budget is a tough job. Unfortunately, it is difficult to really know what things are going to cost until you've gone through the entire process of planning. One way to avoid unexpected costs is to be aware of the kinds of *hidden* costs you can expect. For example, does the wall you want to move contain heating runs? Will your electrical system need to be completely revamped in order to accomplish the renovation? How difficult will it be to get the needed construction equipment into your yard? My next-door neighbor is having a pool built. Getting a backhoe in his backyard was not the easiest thing in the world to do. The contractor spent a great deal of time measuring and looking for alternative ways to approach the problem.

It makes sense to get your contractor to give a ballpark estimate *before* finalizing plans. Be sure that the contractor does a complete home inspection before he does the estimate. This doesn't guarantee that there won't be surprises, but it can eliminate some of the risk.

SIMPLE BEAUTY

With your heart set on all you've been dreaming of, you now find the budget is more than your pocket can bear. Remember when I spoke earlier about the need to prioritize your preferences? Well, this is the time to get that priority list out and take a close look at it. This can help in deflating that ballooning budget easier. It's time to reevaluate some of the details. The simpler a plan, the less expensive it will be. Fancy extras such as curving walls, expensive molding, and custom-made windows may have to be eliminated. Choosing less expensive alternatives can make it possible to achieve an architecturally satisfying look while still meeting your original goal of creating additional space.

STAGGER THE PROCESS

It may be necessary to do your project in stages. When I finished my basement, I planned in advance for improvements that would come later. I prewired and preplumbed for a kitchenette and roughed out an additional bedroom and closet area. It's wired and studded and insulated, but still shy of drywall and flooring. A single $5 light fixture illuminates the space. (It does meet the building code standard.)

I also planned for but delayed having bookcases and other built-ins finished until I was confident I could afford them. Instead of ceramic tile in my master bath, I used carpeting. I did make sure that all necessary prep work is in place for the time when I can proceed and exchange it for tile.

Take your time. Rome wasn't built in a day.

I also planned for a porch and patio that have now, four years later, finally been completed. It was worth the wait. It allowed me to save and build what I wanted and not settle for just what I could have afforded four years ago.

DIY

Is there a portion of this project that you can do? Oftentimes, it is the labor of a project that raises the price. Demolition, painting, and cleanup are dirty work, but they don't require any special skills. Barter is another way to bring costs in line. Do you have a friend that's a plumber? Maybe your computer guru skills can be offered in barter?

STICK TO THE PLAN

I have been guilty of this. As they were closing up the ceiling in my living room, I suddenly realized how much I would miss the daylight. I immediately told them to stop and add two skylights. Although they were not outrageously expensive, they were not in my original budget. Trust me. Regardless of the scope of your project, you can always think of something else to add to it. My words of advice: resist, resist, and resist!

One of the biggest mistakes homeowners make when remodeling is to add on extras as they proceed through the project.

However, if you find you have made a *mistake* then discuss it immediately with your contractor. It is far easier to fix something while still in the process than later. If you do decide to make a change, get it in writing—that includes a price for the change. Keep a good record of all changes and/or additions. I discovered an interesting charge on my final bill that I could not account for. Fortunately, I kept good records and my builder is a great guy. It was an honest error on his part.

DON'T OVERDO IT

Amid all the excitement, it is easy to get a little too carried away. Before you know it, you've turned a reasonable renovation into monstrous proportions. Go back to your original list of priorities and get back to basics. Be sure the proposed plan is consistent in detail and character with the rest of your home. The goal is to complement the original structure, not outdo it or make it outdated. Too often, when the renovation is finished, you realize that the rest of the house now looks awful. As a result, you now need to spend even more money to make it look cohesive. I recommend, at the very least, to plan on painting any adjoining rooms to the renovation project. This should be figured into your original budget. Discuss also the impact changes will have on surrounding spaces. Particularly consider how old and new floors will intersect. Be sure the contractor has planned for and priced any necessary transitions from one floor to another.

> *Classic is always the better choice, whether it is classic contemporary, classic French, or classic traditional, etc. Don't let your home be "planned for obsolescence."*

Review the bids and specs. Be sure you understand all aspects and details. Take the time to shop for and know each specific product. Be sure you are comparing apples to apples. Not all windows, carpets, tiles, hardware, or molding are the same quality. Confirm that the color you want is available. Be sure that there is not a difference from one color to the next. Often *designer* colors are more expensive. Be sure your bid reflects this.

Stay in style. Be sure that your new look won't end up looking out-of-date or faddish. Just because something is "trendy" doesn't mean it will have long-lasting power.

Words of Wisdom

❀ One of the biggest advantages to a remodeling project is the opportunity to "clean" house. Unclutter your spaces—take stock and have a garage sale!

❀ There is no better opportunity to sort through the "stuff" you've gathered over the years than when you are preparing to paint, decorate, or remodel.

❀ Keep a log of all your household projects, especially of things that need annual maintenance or attention. Keeping a home-maintenance schedule handy is also a good way to remember to do such things as the monthly cleaning of the air filter on your HVAC unit. I keep my schedule on a bulletin board on the stair wall leading to the basement.

❀ Remember when budgeting to include new furniture and furnishings. A new space will not meet your expectations if you never finish it. Furniture, draperies, accessories, and accent items are all part of the project and therefore the budget.

❀ The national average for the cost of a family room addition is $32,558 for a 16-by-25-foot room.

❀ If you have a bungalow-style home, try emphasizing its arts and crafts appeal with natural wood and hand-crafted tiles.

❀ To make a room seem larger, try removing draperies so more light gets in. Add a skylight. Replace an old acoustic ceiling with a plain white ceiling, which will reflect more light.

❀ A garage bay can become quarters for: an in-law, nursery, home gym, or a rainy day playroom.

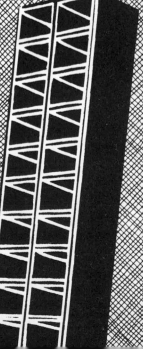
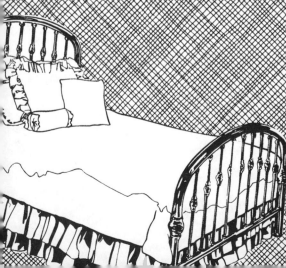

8

MASTER SUITES AND OTHER BEDROOMS

The Master Suite

etreat, haven, refuge, dream palace, nest, or whatever your favorite term—the master bedroom has become one of the most important rooms in the home. Unlike the "public" spaces, this is a space dedicated to *you*. In the Home of the Future (a project built by cooperating builders and suppliers to test-market new products), near Dallas, Texas, this *personal domain* has six separate living spaces within the bedroom: a vestibule, master bedroom, circular sitting room, walk-in closet, master bathroom or spa-lounge (which has three separate spaces itself), and an atrium bridge. In addition it has a gas fireplace and a vaulted skylight ceiling, which is more like an observation station to the sky. Apparently we expect limitless possibilities from our master suite of the future.

I am convinced that it will soon be possible to "live" completely within our bedroom cocoon. Many of the larger homes have incorporated "kitchenettes" to make it easy to get a snack while relaxing in the comfort of your personal hideaway. Our bedrooms are wired for sound, intercom, safety, video, and the Internet. We have virtual ocean-wave sound machines to lull us to sleep and custom-orchestrated music to wake us up. It makes me wonder what the future master suite will be like. I wonder because there are some conflicting opinions as to the general trend of housing for the next millennium. William McDonough, dean of the school of architecture at the University of Virginia, believes the overall "footprint" or size of homes will be reduced. His office, based in Charlottesville, Virginia, is responsible for the master plan of Coffee Creek, Indiana. This *community* of the future has been planned as a pedestrian-friendly, environmentally responsible neighborhood that integrates nature and housing. McDonough believes that our increased size of housing (nearly 25 percent in the past twenty-five years) is wasteful. As a result, the three-bedroom home designed for Coffee Creek is only 1,788 square feet, with an additional one-story studio adding another 384 square feet.

I guess as long as there are people, there will be differing opinions. So what's your preference? That is really the only opinion that matters, especially in such a private space as your bedroom. As a designer, I prefer to work on the master bedroom last. I prefer to get to know the client in the more public spaces first. By doing so, I feel more confident in helping them make choices that are right for them. I have some clients who just want to sleep in their bedroom. And I have others that insist that a TV, ironing board, exercise equipment, laundry, library, and sewing center be included in the master suite.

Here are some thoughts to consider while making your wish list during the planning process:

Closets and Built-ins

❀ I have noticed that there are two different approaches or philosophies when it comes to the basics. There are those (and I am one of them) who prefer a minimalist approach to furniture and choose to use *built-ins* for storing clothing rather than using dressers, wardrobes, or chests. Then there are those who love the look of bedroom furniture and therefore prefer to use it for clothing storage. Your preference will affect and determine how much space should be delegated to each specific area of the master suite.

My reason for preferring built-ins versus freestanding furniture is simple. I got tired of shoving clothing into drawers and having them end up a wrinkled mess. I also got tired of having to take everything out just to find what I needed. I finally solved this dilemma in 1981. I discovered (what was then a new idea) wire baskets and used them to organize my closet. By eliminating the need for furniture for storing clothing, I am able to choose furniture for other purposes. When I fell in love with a beautiful armoire, I chose it to house the television and some exercise equipment in the bedroom area, rather than clothing.

BEDROOM LAYOUT

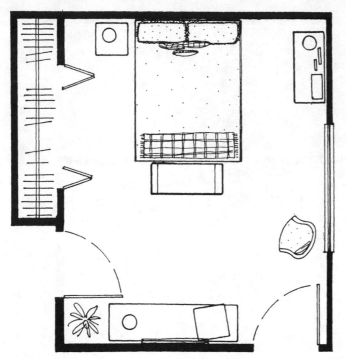

BEDROOM LAYOUT

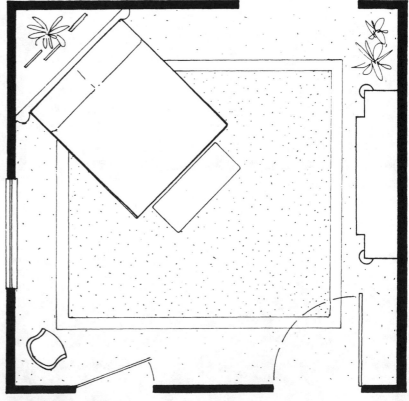

BEDROOM LAYOUT

My armoire is the second most important piece in the room, adding warmth and character. It is second in importance because I believe the *bed* should be the focal point of the bedroom. When you walk into the bedroom, you should be facing the "foot" of the bed, if at all possible. This designates its place of importance.

❧ Since that time, closet organizers have become a standard. You can spend as little or as much as you want to organize your closet/ dressing room. In fact, I have seen a trend toward "dressing rooms" again. It is interesting how often we revert back to older tried-and-true ideas. Recently I turned an extra bedroom into a "closet" for a client. We had the room completely outfitted with custom drawers, shelves, scarf and shoe shelves, and hanging spaces to accommodate a variety of dress lengths. We decorated and accessorized the space, including designing beautiful window treatments.

If you decide to have a dressing room or large walk-in closet, it will eliminate the need for unnecessary furniture in the bedroom portion of the master suite. This gives you an opportunity for many other practical options. In designing my own bedroom, I purposely moved the entry door 14 inches out from the corner. This 14-inch-by-9-foot space allowed me to build in a bookcase with storage cabinets below. There is also room for a sitting "corner" as a result of not using a chest of drawers. Built-in bookshelves at the end of a long room, or even surrounding windows, can give you the library you've always dreamed of having with very little additional expense.

❧ Your budget will also be affected by whether you choose built-ins versus furniture. When planning, take this into consideration. You may want to go window-shopping for furniture as part of the planning process. By selecting furniture in advance, you will be able to accommodate the budget and the specific dimensions and square footage necessary for its placement. After the cost of construction, new furniture usually represents the next biggest expense when remodeling.

❧ Most homeowners prefer a king-size bed in the master bedroom. If at all possible, design your plans to meet this preference, even if you have no intention of using a king-size bed yourself.

Sleepless in Seattle

❧ The preferred location of the master bedroom is the rear of the house because it is generally quieter than the street side.

WINDOW SEAT WITH BOOKSHELVES

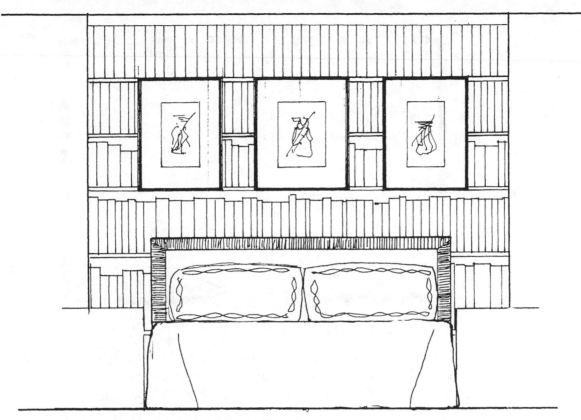

BUILT-IN BOOKSHELF BEHIND BED

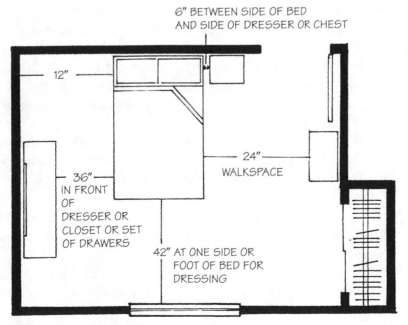

6" BETWEEN SIDE OF BED
AND SIDE OF DRESSER OR CHEST

12"

24"
WALKSPACE

36"
IN FRONT
OF
DRESSER OR
CLOSET OR SET
OF DRAWERS

42" AT ONE SIDE OR
FOOT OF BED FOR
DRESSING

MINIMUM CLEARANCE IN BEDROOM

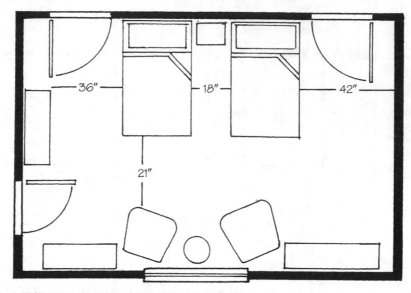

MINIMUM CLEARANCE FOR TWO BEDS

❀ A recent study indicated that the quality of our sleep can be negatively affected by too much light in a bedroom, particularly at sunrise. Too much light keeps you from reaching the optimum sleep level for deep rest. I have always had a difficult time sleeping. I guess you could say I'm an insomniac. As a result, I found this study particularly interesting because I have always used only a minimal amount of window covering on my bedroom windows. For example, currently I have a very large window with an arch-top window above. I have virtually no window treatment or covering on the upper window. I like waking up to a room filled with sunlight. According to this study, this may be the very reason I have such difficulty with sleep. So as you plan for windows in your new bedroom, also remember to plan for their window treatments.

❀ Allergies, particularly to dust mites, are on the rise. Most experts believe it is the result of the increased use of wall-to-wall carpeting. The bedroom is one of the worst allergy-provoking rooms in the house. All the fabrics, feathers, and bed linens contribute to the cause. It is definitely worth including an air-filtration system in your budget to help reduce existing allergies or prevent any future ones from developing.

❀ In working with clients over the years, I have found that there are two basic approaches to color and ambiance for the master bedroom. The preference is either for a soft blush color scheme, or for the cocooning affect of an exotic deep depth of color. Regardless of your own choice, the key is to create a room that brings you to a place of stress-free relaxation. When you enter your bedroom, it should immediately calm the senses. A frazzled, frenzied color and/or style is not appropriate for this haven of rest.

❀ A trend that I have seen over the past ten years is to place the walk-in closet too near the bathroom, which results in a mildew problem. Shoes, handbags, and even clothing can be affected with mildew. Once this problem starts, it is very difficult to stop. I have also seen this happen when a new addition is built on slab (versus a foundation with a basement). The concrete slab can act like a "wick" in areas that have a high water table. It pays to talk with neighbors. Find out if anyone nearby has had a water or mildew problem such as this.

❀ Not everyone has the luxury of a master suite. If this is your situation, don't fret. There is still plenty that you can do to improve the

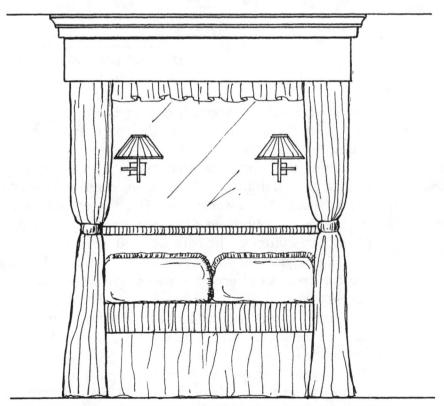

BED WITH MIRROR BEHIND AND CEILING-MOUNTED CANOPY

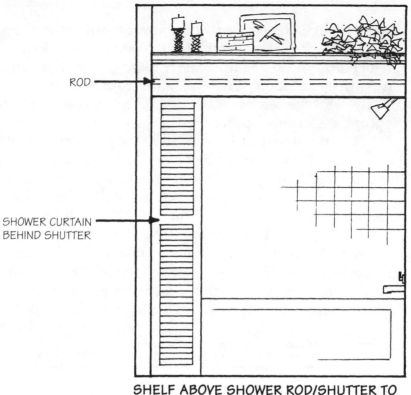

ROD

SHOWER CURTAIN
BEHIND SHUTTER

**SHELF ABOVE SHOWER ROD/SHUTTER TO
HIDE SHOWER CURTAIN**

appearance and functionality of your bedroom. Consider using mirrors to enlarge the overall affect of the space. For one client, we mirrored the wall behind the bed. To create the look of a canopy without overwhelming the space, we used molding on the ceiling from which we hung a pleated drapery valance. Lamps were mounted through the mirror, giving the bed a warm glow. The master bathroom, which was from the 1970s, was updated with new cabinetry and tile. We used a tall plantation shutter at the bathtub to hide the shower curtain behind.

❀ Do you and your mate share the same sleeping schedule? This too can have an effect on how you plan the master suite. It may be necessary to subdivide the space, using French doors to keep the night owl from keeping you awake. If you are knocking down a wall between adjacent bedrooms to create a master suite, the use of French doors can be ideal.

❀ My night owl husband is a nighttime reader. I installed a tiny reading lamp at his side of the bed. This allows him to comfortably see without keeping me awake.

❀ If you like to watch TV in bed and your spouse doesn't, consider purchasing a television with a headset.

❀ Get wired! Just how far do you want to go with wiring options for your bedroom? Computer networks, video and sound, security, phone, lighting, Internet, and intercom all need to be planned and budgeted in advance. Lighting, security, climate control, fire and safety and entertainment systems are all integrated. It is, for example, possible to prearrange your lighting to turn on and adjust to 20 percent of full illumination for the first twenty minutes of the day while simultaneously motorized blinds open and soft music plays. Or you may choose a complete entertainment system, including complete satellite hookup, multiaudio zones, and speakers that drop down from the ceiling! Even if you don't go to such extremes, it does make sense to at least take into consideration such things as security systems and interior as well as exterior lighting control. I always make sure to include the ability to control turning on exterior spotlights and general interior lighting from bedside. That way, if you do hear an unusual noise in the middle of the night, you can at least see what's going on.

Beyond the high-tech world, the important thing is how your room makes you feel. Your bedroom should be the one place where personal reflection is most accepted. Whether you're refined, exotic, traditional, monochromatic, neoclassic, country, or contemporary, it should make you feel relaxed and special.

Your Children's Rooms

The key to this section is recognizing that even young children have a need (and I think a right) to be involved in designing and decorating their own spaces. Children who *like* their room tend to *care* for their room. Allowing them to have a sense of ownership and pride goes a long way in instilling respect.

At a recent speaking engagement, one mom brought along her eight-year-old son. She had expertly equipped him with reading, playing, and eating materials to keep him adequately amused throughout the evening. They were some of the first guests to arrive. I introduced myself and had a very engaging conversation with this eight-year-old gentleman. I asked him if he would be willing to participate in a little test—what I call my Ink Blot Test. He agreed. At the appropriate moment of the evening he graciously came to the front and proceeded to take this style color test. We soon discovered that his preference was for green and highly contrasted schemes. What I also discovered was that "mom" had made all the decisions for decorating his bedroom. It was "blue" not green. And he eloquently explained all the other things about "his" room that mom had done that he did not like. Everyone laughed. But the point is, he knew what he wanted. So why not accommodate him? Of course, you have to be reasonable and there are limits to creative expression. But some of my most rewarding design experiences have been with children.

When I say "children" I don't just mean young ones. Preteens, teens, and college-age children are also part of this discussion.

Allow your child to be involved not only in the design but also in the actual process of decorating. I have involved preteens and teens in many different painting techniques. Ragging, sponging, even trompe l'oeil painting (for example, my niece painted beautiful clouds and angels on her wall and ceiling). Paint is inexpensive and easily changed. Use this opportunity for playful ingenuity.

Most parents are cognizant of the fact that children grow. The challenge given to me as a designer is to find a way of making a "plan" last as long as possible through all the growing stages. The ideal situation would be to redo a child's room at each major interval: infant, toddler, grade school, and teen. Realistically, you will probably redo their room only twice in their lifetime. So whatever you do must work for a long period of time. The basic necessities are: a bed, a play area, a work surface, and storage for clothing and other belongings. Of course, in a child's mind, all of these things are multipurpose items. Climbing on, under, and around them is to be expected. Therefore, be sure everything is "nailed" down. Stabilize anything that has any potential for falling down or being knocked down.

One problem most often expressed by clients with regard to children's rooms is organizational storage. Whether it's stuffed animals, toys, clothing, or just junk, kids manage to collect stuff! I've often been called in to do "animal management"—stuffed animals that is. Compounding the situation, most children's bedrooms are rather small. So what can you do?

My approach is to build in, up, and around. Add shelving/shelves, cabinets, and nooks wherever you can. There is a lot of unused space near the ceiling, at the top of the closet and around windows. Bins, baskets, boxes, and tins can really help get things organized. Multilevel building can also add interest as well as functionality to a bedroom. Don't forget about that precious space underneath the bed. Purchase storage units designed specifically for this purpose. You will gain not only much needed storage, but it also keeps unwanted things from finding their way under the bed.

Try a "seasonal" approach to clothing and athletic accessories. Put out-of-season things up out of reach, switching as seasons change. Rotate toys—it keeps children interested longer. (My mom regularly packed up a bunch of toys and put them out of sight. A few months

SHELVING AROUND CHILD'S ROOM

BOOKCASES IN BEDROOM

later, she would recycle them. This kept us interested longer and made keeping things neat easier.) Recognize and acknowledge your child's preferences and style. Realize that teens have an innate desire to decorate their room with posters. Let them! Forget fancy wallpaper, you'll never see it. Instead cover an entire wall with corkboard and let them go at it. Corkboard can be painted to match the room. But honestly, you probably won't be able to see it anyway, so why bother?

An observation I have made is that people are innately messy *or* organized. Unfortunately, an organized child can be highly influenced by a messy parent. If your own organizational skills leave something to be desired, then expect that this will affect how your child keeps his or her room and personal items.

Make the room easy to maintain and clean. Choose materials that are durable and practical. When choosing carpeting, select a low-pile or loop construction. They require less vacuuming to keep them looking good. Choose colors that are good at hiding soil and lint. Carpeting that is either too dark or too light can be a problem. One will show dark stains and lint and the other light stains and lint. If you expect your children to make their own bed, then make it easy to do so. Forget all the complicated bed treatments. A throw or comforter is relatively easy to manage.

If you expect your child to actually play and work (doing homework, for example) in their room, then make it both inviting and functional. Be sure that there is sufficient light and a good working space.

The older a child becomes, the more time they will spend in their rooms and the more they will accumulate. Sound systems, telephones, computers, and TVs all too soon find their way into their rooms. To make life easier all around, try to designate a "den" for teens—whether it's a basement rec room, a barely used guest room, or simply a large hallway. Having a space of their own can make life more pleasant for everyone in the house. Especially when all of their friends show up to watch "Fright Night" movies. Do you really want them in the middle of your newly furnished family room?

Depending on your particular situation, your guest room is either used often or is basically a museum room that nearly never gets used. If it's a museum room, then consider making it a multipurpose room. It can be the perfect place for a den, a craft/hobby room, library, or computer/office space. Whatever your preference is, don't let this valuable space go to waste.

A guest room obviously needs a bed or beds to make it functional. Two twin beds generally make more sense than one queen- or king-size bed. That way, regardless of gender or affiliation, two people can easily use the space.

In my opinion, if at all possible, do not use a sleep sofa. Most of them are impossible to sleep on and horribly hard and uncomfortable to sit on which makes them virtually useless for any purpose. As a guest, I would much rather sleep on a good quality air mattress than on a sleep sofa.

New Spaces

The most popular way to add a master suite or children's dormitory suite is over the garage. It is the easiest way to give an older home modern amenities. This can also be an opportunity to add a second interior stairway to the home—another old idea that's become popular again, particularly in new construction. By adding a second stairway, you can also create some interesting smaller spaces. An office hideaway can be tucked into a loft at the second-floor landing, or in an enlarged lower level landing of the new stairway.

Skylights and large expanses of windows will not only illuminate the space but add to the overall spaciousness as well. Be careful though in the planning, selection, and placement of windows in the new master suite. A few years ago, I received a phone call from a family that had just completed the construction of a new master suite. They suddenly realized that their beautiful, new, seven-foot, arch-topped window gave their neighbor a front-row view directly into their bed! Something they obviously had not planned. A mere driveway separated their home from their neighbor's—they were virtually window to window. In their minds, they had anticipated daylight streaming into their bedroom. The reality was that they had to cover nearly all of the window with privacy shades.

Window placement is critical not only for privacy but also for placement of furniture, artwork, and the aesthetics of the interior and exterior of a home. Style and proportion of windows are also important to creating an addition or renovation that looks as though it belongs with the rest of the home.

Ultimately, what separates a master suite from a bedroom is not the sleeping space itself, but all the additional spaces—the luxury of privacy, a bedchamber, a master bath, a dressing room, and well-organized closet spaces. It provides a combination of spatial qualities and a maximum amount of functional luxury.

Words of Wisdom

❀ Whenever you increase the value of your home, be sure you also increase your homeowner's insurance policy. Your home should be insured for at least 80 percent of the actual cost of rebuilding. Since furniture and other possessions are usually depreciated in value, it is worth paying extra for a *replacement value* policy. This usually costs only 10 to 15 percent more. One of my staff recently had a very unfortunate experience. While relocating, her furniture was in storage at a moving company. There was a fire, and everything she owned was lost. Fortunately she had a replacement-value policy and was able to purchase new furniture of equal value.

❀ Every room of your home should be comfortable, functional, and inviting.

❀ I don't think you can ever have enough closet space. Use every little nook and cranny to give yourself more places to organize your personal treasures.

❀ By using a limited number of patterns and prints in your bedroom, the easier it is to create a room that calms the spirit.

❀ Heating and air-conditioning considerations need to be addressed early in the planning process of a new bedroom suite. Your existing HVAC (heating, ventilation, and air-conditioning) system will need to be evaluated to determine if it has the capacity to properly handle the additional square footage. In most cases, you will need to supplement it with a second unit.

9

BASEMENTS AND ATTICS

Basements: Work or Play—Realistic Expectations

As a designer, I have finished and remodeled a lot of basements. It seems that no matter how big a home is, we still want more space—space for playing, working, storing, exercising, watching movies, doing laundry, or catching up on hobbies. And that big empty area below the rest of the house seems like the ideal place to gain additional square footage. Or is it? This will depend on a couple of very important factors: dryness, accessibility, and ceiling height. These are the first three things to check out before making the decision to renovate your basement.

Moisture and dampness, in most cases, can be corrected. It's just a matter of how much effort, money, and work will be necessary. The first thing to check, is to be sure that landscaping and soil slopes down and away from the house. In some cases, it may be necessary to install a trench or underground drainpipe around the house to funnel off water. Installing a dehumidifier is almost always necessary in a basement. Venting your HVAC (heating, ventilation, and air-conditioning) into storage areas of the basement can sometimes help reduce moisture. Patching cracks and crevices, and then sealing them with a cement sealer, will also be helpful. In addition, I always recommend installing a sump pump for those unexpected downpours. The last thing you want is a storm to flood your just renovated rec room.

Access and ceiling height can affect how you will ultimately use your finished basement space. Building codes for municipalities differ from one town to the next, but most are specific with regard to usage

The latest trend in new construction regarding basements is to design them as an extended part of the rest of the living space rather than as a secondary afterthought. Open stairways with details and qualities consistent with the design and finish of the rest of the home are becoming standard. You no longer access them through a closed door off a small hallway or kitchen. Instead, grand, wide, open-spaced stairways and landings now lead to a home office, exercise/spa, family entertainment center, and guest bedroom suites.

for such things as bedrooms, with a minimum of 7½ feet of finished ceiling height. In some cases, it may be possible to lower the floor (by excavating it downward) to gain the needed height. This can be expensive and may require reinforcing the foundation.

Accessibility and/or a fire exit will also be required. A daylight/walkout or above-ground exposure basement is not usually a problem because it will allow for easy exiting through exterior doors or large windows. However, if you have a standard below-grade level basement, be sure to check local zoning regulations for access requirements.

FINISHING OFF A BASEMENT . . .

... TO BLEND WITH REST OF HOME

Budgets

Budgeting the basement renovation project will completely depend upon your intended use. The two most popular uses for basement space is to build a rec room (usually a teen retreat) or a home office. The national average for a 12-by-12-foot home office, according to *Remodeling* Magazine's Cost & Value report, is $8,179. This included custom-built and installed cabinets with desk, computer workstations, overhead storage, and 20 feet of laminate counter space. It also included rewiring of the room for computer, fax machine, and other electronic equipment such as telephone and cable. Drywall interior and commercial-grade carpeting finished the office.

Finishing my own basement cost me an average of $12/square foot for the basics. This budget was basically for finishing walls with drywall, and a commercial grade carpeting for the floor. I chose not to use a floating floor or padding underneath the carpet. This budget did, however, include a full bath and subdivision of the basement space to create a guest bedroom/office, a rec room (which included preplumbing and wiring for a future kitchenette), and a roughed out plan for an additional bedroom.

The Floor Plan

Designing a floor plan to fit your needs is key to a successful project. The basic considerations for any basement redo are:

❀ Be sure it's dry. If there is any suspicion of a serious water problem, call a professional waterproofing specialist.

❀ Consider and plan to accommodate a maximum of activities. Divide the space into specific areas for video watching, relaxing, exercising, and doing crafts.

❀ Create interest in your plan to compensate for the narrowness and darkness of most basements. By varying wall and ceiling planes you can create a visual treat for the eyes and still accomplish necessary divisions. Half walls or even a change in ceiling height or definition can define spaces in unique ways.

❀ Decide on your furniture plan as you are designing the rest of the space. Furniture can actually be used to define spaces by its placement.

❀ I love angles; they create the illusion of more space. An angled wall or placement of furniture can help alleviate the narrow feel that most basements are prone to.

❀ Lighting is particularly important for basements with few or small windows. I prefer to use a combination of fixtures to get the maximum benefit. A mixture of recessed ceiling fixtures and wall-mounted sconces can add not only light but interest as well. Fluorescent ceiling fixtures are a practical and inexpensive way to add a great amount of light to work or task areas. Give special attention to eliminating dark corners.

❀ Choose your finish materials to create visual interest as well as for their practicality. By mixing elements such as wood,

Lighting, lighting, lighting! This is one of the most important considerations for creating a space that encourages use. Nobody wants to go down into a dark dungeon of a basement.

fabric, texture, and metal you can add dimensional value that appeals to your sense of comfort and invites you into the space.

❀ Carefully plan your color palette. Color affects us on so many levels. Bright, cheery colors are key to creating a space that improves your mood rather than the opposite. Color is a more important element to the overall plan when there isn't much natural light available. Contrasting colors can also go a long way in creating architectural or visual interest in an otherwise dull space.

❀ Plan storage space and use every available inch. In my experience, one of the most efficient ways to get the most use out of a basement space, whether for rec room, home office, exercise room, or craft/hobby room is to use built-in cabinetry. Designing cabinetry to fit the need will allow you to get maximum use in the most efficient and decorative manner. Storage, organization, and utilization are the key to creating a space that not only functions well but is also pleasing to the eye. Don't let the words *built-in* scare you. I recommend fastening cabinets to the wall temporarily. There are two advantages to this: first, if the cabinet is anchored to the wall, it is now a home improvement (a simple "L" bracket can work), and in most states you will not have to pay sales tax; second, it gives you and the next homeowner flexibility. The key to this flexibility is to let the cabinet builder/installer know in advance your intentions. I have built, installed, and moved many cabinets. Generally it will cost only a fraction more to build cabinets that can be moved. Obviously, not all built-in spaces are conducive to moving—understair closets or computer stations, built-in closets, odd-shaped but handy alcoves for storing sports equipment or craft supplies can all make life much neater but generally they will be permanent installations.

❀ Flooring choices are important aspects not only for their decorative value but also for their ease of maintenance. Recently, in replacing flooring in an existing finished basement, we changed from carpeting to laminate flooring. The carpeting had always been a problem—five children all wanting to play indoors in winter had played havoc with the old carpeting. The new laminate flooring is beautiful and durable, making life easier for everyone, particularly Mom.

Most basements have concrete slab floors. You can, if they are level, install carpeting, or vinyl/resilient flooring directly. You can also choose to stain the concrete, creating a tile or stone look. For better

SHELF AROUND PLAYROOM

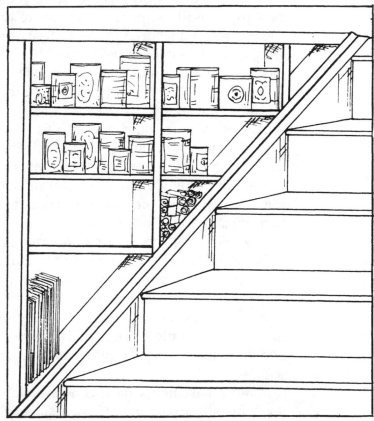

BASEMENT-STEP STORAGE

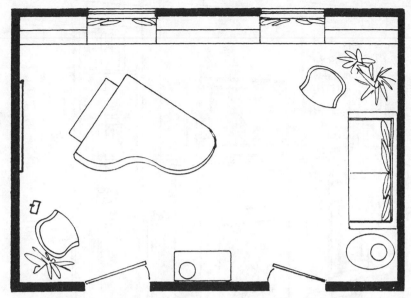

FLOOR PLAN SHOWING BOOKCASES BUILT AROUND WINDOWS

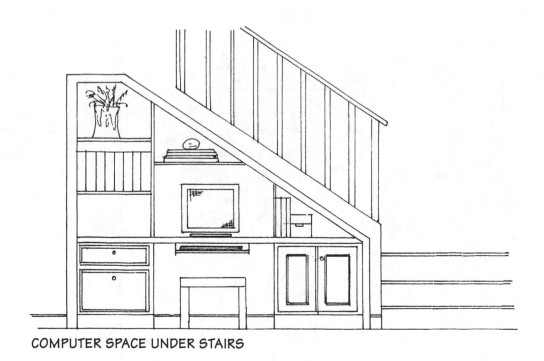

COMPUTER SPACE UNDER STAIRS

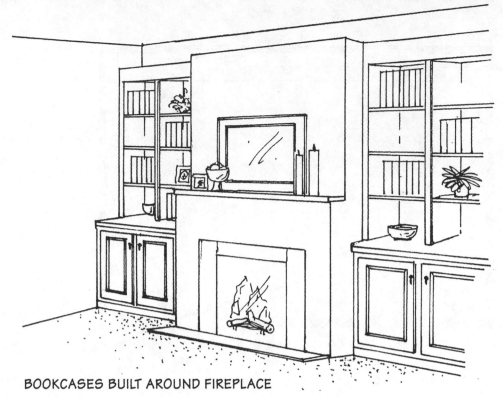

BOOKCASES BUILT AROUND FIREPLACE

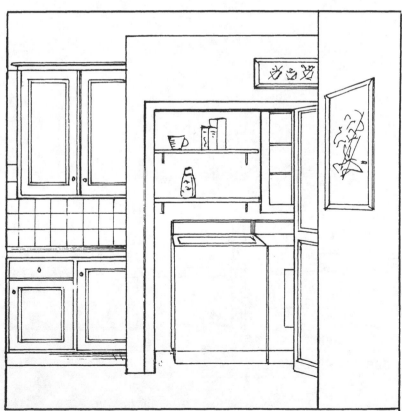

LAUNDRY ROOM IN KITCHEN CLOSET

insulation, a plywood subfloor can be framed and installed over the concrete before installing a finish material.

❀ Ceilings can add sound insulation to your basement, which can make a big difference to keeping loud noises such as loud music contained. It's also a benefit for a quieter home-office space. If possible, use resilient padding to separate pipes from wood frames. Provide air chambers to eliminate water hammer, which is caused when you quickly shut off a faucet. Choose sound-absorbing surfaces. Armstrong World Industries (800-233-3823) offers 2-by-2-foot acoustic ceiling panels that have a step-edged detail or look like embossed or molded plaster. These are wonderful for blocking noise and keeping it from invading upstairs. They can give your basement as STC of about 35.

If many basements plumbing pipes and steel posts that hold up girders can be a problem. I recommend simply boxing them in with plywood and finishing them off to match the ceiling and wall. I do not like to draw attention to pipes and posts with techniques such as making them look like wooden beams. This draws your attention and lowers the overall appearance of the ceiling. By keeping them the same color as the rest of the ceiling, the eye will be fooled into overlooking them.

❀ Heating and air-conditioning needs to be considered. In many cases the surrounding ground tends to moderate the temperature of the basement. Using a louvered screen or vent in the furnace room can provide radiant heat to the rest of the basement. Heating that is forced warm air can usually be extended from the main system. However, this can affect the overall efficiency of the system. In my own home, keeping the rest of the house cool in summer with air-conditioning makes the basement too cool.

Another item to consider for "healthy" air is a carbon monoxide detector. Carbon monoxide can escape from any fuel-burning appliance, such as a furnace, water heater, fireplace, woodstove, or space heater. Improper ventilation can also cause a build-up of this poisonous gas.

Since it is a colorless, odorless, and tasteless toxic gas, we are often not aware of a problem until it has already had a serious affect on our health. The simple solution is to install a UL approved detector.

Home Office—Special Considerations

Statistics indicate that the return on investment for a home office has been climbing every year since 1994. Real estate agents expect an office remodel to return 69 percent of its cost this year. This trend, according to *Remodeling* Magazine's Cost & Value report is strong in every region. The home office is a hot item! An estimated fifty-two million Americans are now conducting at least some of their work from their home via computer. This mobile working style is changing the shape of the American home. Developers in upscale condominiums and homes are designing offices placed directly off the entryway that are completely separated from upstairs living quarters. New developments nationwide are marketing high-speed telecommunications, separate offices, or both. Even Disney Company's planned Florida Celebration development allows some mixed business use with residential.

The home office is here to stay. Telecommuting has changed the way we work.

❀ Unfortunately, not all builders are up to speed on the requirements and specifications necessary for this brave new world. My home was built in 1994. As soon as I moved into my new home office, I knew I had a problem—cross talk on the wires! In other words, I could hear conversations from one phone line to the next. The electrician who wired my home had no idea that he needed to upgrade the wire in order to allow for the use of many different electronic devices without creating a problem. Unfortunately, since the basement had been "finished" it was now a virtual mess to redo the wiring. After a lot of discussion and work, the problem was solved.

Take this as a warning—be sure the guy you're talking to knows what he's talking about. High-speed T-1 telecom lines will enable you to quickly download files from the Internet or corporate databases without a cross talk problem.

One of the most important aspects of creating a home office is to make sure it has high-tech wiring to meet the demands of a high-tech world.

❀ Choose appropriate furnishings. If you are only going to use your home office for part-time work and don't expect to do a lot of desk or computer work, then an inexpensive workstation can be sufficient. However, if you're planning on really working from home, don't be a cheapskate, you will only pay for it in the end with real pain. Computer "tennis elbow" and carpel tunnel wrist are serious problems. I know, I have spent nearly a year in physical therapy for "tennis elbow." I finally invested in a properly designed, ergometrically correct desk/computer station, with a multiadjustable chair, which allows me to keep my elbow height consistent with the computer keyboard and mouse. In addition, because I'm a petite five foot two, I have a footrest under my desk to properly support my legs.

When selecting a desk, look for one that is at least 20 inches deep to accommodate a computer monitor. The best height for a computer keyboard is 26 inches. Be sure you don't have to tilt your head back or forward to view the screen. It should be about 14 to 20 inches from your eyes.

❀ Plan for at least two types of options: general lighting for allover illumination, and task lighting for your work surfaces. Be sure you can keep glare away from your monitor screen. Proper window treatments that help control natural light are essential. Since the monitor screen is self back lit, it is not generally necessary to add additional illumination to it. Use a task specific lamp to illuminate only your work surface.

❀ Control static electricity and dust. These are two of the biggest bothers for computers.

❀ Get organized! I outfitted a closet in my home office with a tall vertical file as well as wire shelving. This is a great way to get storage out of sight. A window seat can also double as a file or other storage.

Home Theater—Lights, Camera, Action!

Over the past five years, a new era has developed—home theater. Our infatuation with being able whenever we choose to rent and watch movies without commercial interruption has turned into a love affair. A home theater is more than just buying a big-screen TV. There are two specifically different aspects to home theater. First is the atmosphere and design of the room, complete with plush movie seats, special lighting, curtains, and a lobby with a popcorn machine. Second is the incredibly complicated world of digital-surround technology.

A home theater can be anything from a room with a hi-fi VCR and a stereo TV to a Dolby-Pro-Logic receiver with surround sound speakers, subwoofers, laser disc player, DVD player, and a graphic-grade front projector! Whew . . . I needed an extra breath to get all that out. The point is, home theater is a big and growing industry. Technologically, it will be very difficult to predict just how far all this will go, which is a bit scary, considering the cost of all this personal entertainment.

The basement is one of the more popular places to build such a center. The reason for this choice is that it is one space that is fairly easy to isolate in terms of containing the sound. This is an important issue because home theaters usually mean very loud, very distinct, megaprojecting, and reverberating sound. Being able to contain this sound (noise) within its designated space is very important to the sanity and sleep of those who are *not* watching (and listening).

The professionals' recommended choice for sound-dampening products is fiberglass. Some experts suggest lining walls, ceiling, and floor with fiberglass. Owens Corning's Acoustic Laboratory has a reversible fiberglass-and-gypsum panel system. Each panel incorporates a reflective and an absorptive surface. This allows you to create an environment free of unwanted sound reflections. This system is designed specifically to be retrofitted into existing spaces, or used in new construction, making it an ideal choice. The finishing touch is an extra-wide seamed fabric that is stretched and anchored at the room's perimeter, hiding all the background work. The panels themselves are adaptable to unusually tall spaces, as well as awkwardly shaped spaces.

A few years ago, only the very rich, usually from Hollywood, would even think of having a home-screening room. Today, there are countless specialty retailers providing design, installation, and education to "Joe" homeowner. The cost for such a room runs the gamut from a few thousand dollars to many thousands of dollars. A home theater room with acoustically sound walls, plush seats (seating for six), curtains, and a minilobby can be done for around $8,000 to $10,000, excluding the cost for video and stereo equipment. Equipment pricing is impossible to average. There are so many options, all

HOME THEATER

HOME VIDEO BUILT-INS

of which are way beyond my electronic and technical ability. I have seen some quotes as high as $50,000. This usually consists of a data-grade or graphics-grade *front projector*, with a couple of line-doublers or triplers or quadruplers, as well as a high-precision *screen*. The *speakers* should be THX certified (a set of standards by which laser discs, video-tapes, and home theater equipment is made), with at least two sub-woofers, converting to Dolby Digital format. The speakers are powered by a separate *processor* and *amplifiers* for each speaker.

At those prices, you better take a deep breath before you start considering this too seriously. After all, you're just going to be watching movies on a big screen. For $10 or $20 you could go to a real movie theater!

Attics—Upward Possibilities

Although usually dark and confining, with little headroom, an attic can be a wonderful way to expand. The most popular use for an attic is a bedroom, often a master bedroom, with a bath and walk-in closet. The biggest challenges to your dream room are convenient access, and enough headroom not only to make it usable but also to meet building codes.

The first thing to check is whether or not you can change the facade of your home. If, for example, you live in a historic district, or your home is a registered historic building, you may not be able to change the exterior appearance. This can make an attic renovation much more difficult though not necessarily impossible; you just need to get a lot more creative with the space you have. If your home has no restrictions to changing its appearance, then a whole variety of options are available. Simply bumping out a boxed bay can be a great way to create a focal point with tall, geometric windows. Adding dormers or raising the roof can also give you some interesting nooks and crannies.

Before we get too far into planning this upwardly mobile space, let's talk about some of the basics. Ceiling height will definitely be dictated by your local building codes. Before you begin planning, check your local codes. In most cases, you will be required to have $7^1/_2$ feet of ceiling height over at least half of the space. They will also set minimum standards for emergency fire escapes and ventilation. A window can be used to provide a means of emergency escape for attic living

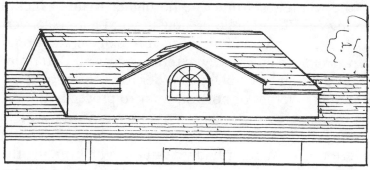

RAISING THE ROOF

ADDING A WINDOW CLERESTORY

BEFORE OF RANCH HOME

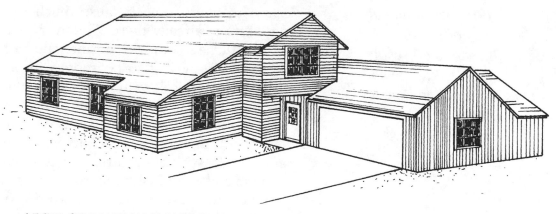

AFTER OF RANCH HOME WITH NEW ADDITIONS

spaces. However, the window sill must be no higher than 44 inches above the floor. In my old Victorian home downtown, I had to convert an existing window to a door for an emergency fire exit. In addition, I had to add a rather elaborate emergency exterior stairway. Determining the upfront costs that can be involved in order to conform to codes will help in determining whether or not you want to proceed. In most cases, some of the more expensive costs to an attic renovation will be in rebuilding interior stairways for sufficient access to meet codes and the emergency exterior exit. Another budget eater can be bringing the floor and rafter structure up to current codes.

An attic conversion usually involves such things as adding insulation, framing, and installing a ceiling, partitions, and knee walls. Additional spaces and alcoves can be added with dormers, new windows, and skylights. Plumbing, electricity, and flooring all need to be considered. Remember, all the building materials that you will need for this project must be able to be moved into the attic space. Accessibility can be a major problem. Often it will be necessary to bring them up from the exterior through a window opening. I have often used a drywall delivery truck with elevator lift to bring many of the building materials up this way. Of course, you could always use the old-fashioned block-and-tackle method. The point is that you need to consider and plan your renovation schedule to accommodate this little stumbling block.

BUDGET

I don't recommend basing all your decisions on resale value only. The bottom line is you have to live in your home. It should be an enjoyable place to be. An extra $500 here or there can go a long way in making it more enjoyable.

An attic bed-and-bathroom conversion is one of the least expensive ways to add space to your home because much of the structure is already there. The average cost of an attic bedroom with dormer and bath is $23,000 (15-by-15-foot bedroom with 5-by-7-foot bath). Resale values vary. Nationwide the average resale value recouped is 86 percent of construction cost. In the West, it is nearly 100 percent. You can almost be guaranteed a 100 percent return on your dollar if you can add a third bedroom to your existing two-bedroom home because two-bedroom homes are nearly obsolete.

HEATING AND VENTILATION

Because hot air rises, heating an attic living space can usually be accomplished by extending the existing system. A floor register installed in an opening cut between the ceiling joists of a room below will allow warm air to rise. The attic stairs can also serve as a passage for warm air. One tip often used is to remove any insulation from the attic floor and insulate the attic ceiling and exterior walls. This will allow easier passage of controlled air through the floor, while maintaining it in the attic. Be aware, however, that this will also increase the transference of noise from one floor to the next. In colder climates I recommend the new high-density batt insulation or foam insulation. Installing ridge vents under the insulation will help control humidity.

You may also choose to install a separate zone heating system for the attic. This can be accomplished with an electric baseboard system. Air-conditioning can be installed with a ventless individual unit. These can be installed indiscreetly and are very effective.

THE PLAN

The best part about an attic suite is the opportunity for interesting architectural elements. The ceiling provides the biggest opportunity for creativity because it's at the top of the house. Opening the ceiling to the roof peak and adding windows that reach to the top will give you a maximum feeling of spaciousness as well as the added bonus of boundless natural light. One of the more ingenious ideas I have seen is to run a continuous skylight along the ridge. This wash of light across the ceiling is a great way to add daylight and still maintain privacy.

Some of the easiest ways to gain space and improve lighting to a top floor of a home is by adding a dormer and/or skylights. A dormer is a small, roofed, almost independent structure that is placed over a new opening in the roof of a home. Because you are not really adding floor space, the cost is not nearly as expensive as other methods. However, depending on the size of the dormer—one, two, or even three-window dormer—you will add headroom and light, and maybe even a place to build in a desk or storage drawers.

Adding a window tower is an option to consider for raising the roof. Architect David Cox did just that to a home in the Washington,

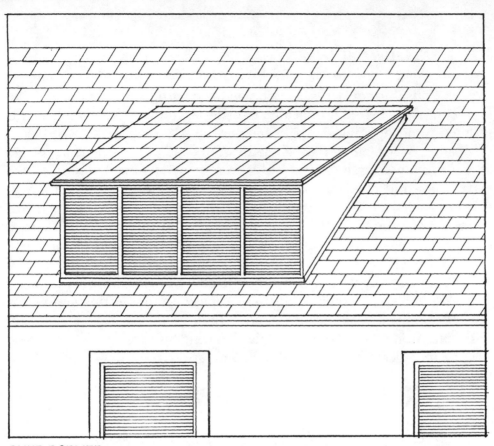

SHED DORMER

DORMER

D.C., area. A 15-square-foot tower with great architectural details of the Arts and Crafts period allowed the new bathroom a spectacular amount of light and volumes of space. He incorporated two antique leaded glass windows into the clerestory, mixing old and new for a wonderful custom-made look.

You can also raise the entire roof to gain needed headroom. Pushing up the ridgeline a mere 4 feet can simplify convoluted ceiling lines to make them make sense.

Another interesting way to allow light from a window to freely flow throughout the space is to use a wall that only reaches partway to the sloping ceiling. The use of semitransparent materials in the stairwell, or walls, such as ribbed glass, can allow for a visually open space that can also be enhanced by the transference of light.

Planning to use spaces for the most appropriate tasks is key to a successful and functional renovation. Obviously you want to use the areas with the highest ceiling spaces for tasks that require headroom: the bathroom and the placement of the bed. You may want to raise a low dormer, or add a dormer, to gain needed headroom for a bath or

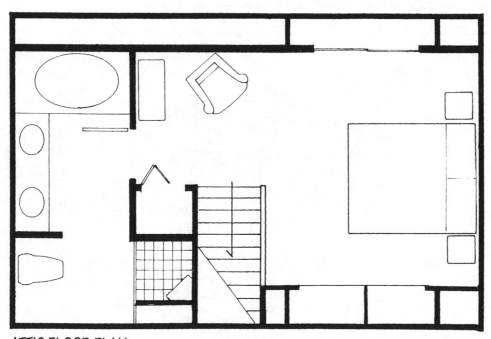

ATTIC FLOOR PLAN

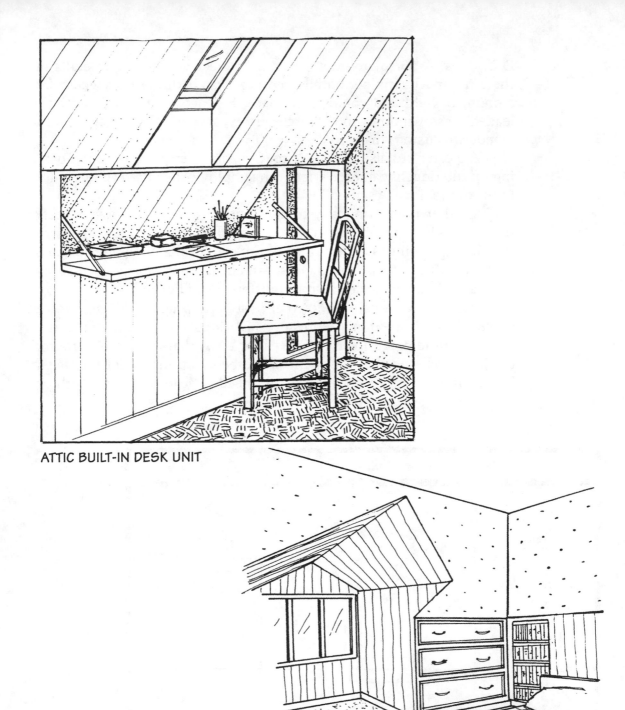

ATTIC BUILT-IN DESK UNIT

ATTIC BUILT-IN STORAGE

shower. Use the shorter spaces, such as knee walls, for built-ins. Dressers, desks, and shelving should all be built-in for maximum use of space. The least expensive way to accomplish this is to use stock cabinetry and build around it.

Keep it light—light colors always work best in an attic redo. Most designers agree that it is the one place where white is right. The interesting angles and ceiling lines bounce the light around, creating a space that makes you feel good just being there.

Words of Wisdom

❀ Before you decide to give up the rush hour commute and telecommute, be sure your local zoning regulations will allow your particular type of business to operate from your home. Recently in our neighborhood, there was quite a dispute when a local lady tried to open a manicure business in her home. Ultimately, she lost.

❀ Just because it's a home office doesn't mean it has to look unprofessional. It's really okay to have file cabinets and a real desk!

❀ Take advantage of a great view from the attic by using large expanses of windows. Usually the view of the trees is great! Do plan a window treatment that will allow you to get a maximum of light and the necessary privacy.

❀ If you have a traditional-style home, a shed dormer to the rear works best. If not, then try a dormer on the *east* side of the room. This will allow you to wake up with wonderful morning sunshine.

❀ An attic space is a small space. Take this into consideration if two of you are sharing this new room. To keep things quite, use acoustic batts in the bath partition walls to muffle morning sounds—particularly if one of you gets up earlier than the other.

❀ *Radiance Low-E Attic & Decking Radiant Barrier* by ChemRex Inc. (800-766-6776) is a coating that can be applied on the plywood decking in attics that will reduce the heat loss in a building by 8 percent to 24 percent in the winter. It also reduces heat gain in the summer by 11 percent to 28 percent. It is available in whites and pastel colors. It is not available in dark colors, because it is the reflective value of the lighter colors that makes it effective. Darker colors absorb more radiant heat. It is applied like paint.

PART THREE

EXTERIOR RENOVATIONS

10

ARCHITECTURAL FEATURES AND EXTERIOR SURFACE ELEMENTS

The History and Design of Your Home

It's time to move outside and take a good look at the exterior of your home. The first question is: What style is it? This may seem like a silly or simple question, but amazingly, many people have no idea. Sometimes it's because a home has been through so many renovations and changes that it's virtually indescribable. Too often I have seen what started out as a wonderful Georgian traditional home be converted into a mess. Don't get me wrong, I am all for people expressing themselves. However, I do think things should make some sense with regard to aesthetics. So here are some basic descriptions of home styles:

❀ *Cape Cod.* One of the most popular styles because of its compact design. These houses are an economical one and a one half stories high. They are very symmetrical, with the entrance centrally located. The roof has a steep gable. The exterior can be any material—brick, wood, or vinyl siding. They are a popular choice for retirement homes because they offer one-floor living, with a first-floor bedroom. The additional space in the upper level can be finished to provide extra square footage.

❀ *Dutch Colonial.* Usually two or two and one-half stories high with a gambrel roof. Gambrel refers to a roof having two slopes on each side. The lower slope is steeper than the upper (flatter) slope. The eaves flare outward. They are traditionally made of shingles or brick. Dutch colonial–style homes are most often found in the mid-Atlantic region.

❀ *Georgian.* One of the most formal of styles, it is a classic two or three stories high, built of either brick or wood siding. It is stately with its rectangular shape and flanking columns at the entry. A multipaned window transom above the front door is repeated with multipaned windows throughout the home. This eighteenth-century beauty usually has two large chimneys that rise high above the roofline at either end of the house. A center-hall foyer with an elegant stairway rising up all levels includes large stairway landings—perfect resting or reading getaways.

❀ *New England Colonial.* An American classic two and a half stories high that is boxier than its Georgian cousin. It has a shingled gabled roof, and is most often built of wood clapboard siding. The windows are double hung, small paned, and originally had working wood shutters on the outside.

❀ *Pueblo.* This southwestern style, often referred to as Santa Fe style, is either adobe brick or framed with stucco finish. The roof is flat with rounded exposed beams. They are usually one or two stories high and usually have a covered or enclosed patio. Mexican terra-cotta tiles usually cover the interior floors as well as the patio floor.

❀ *Ranch.* Originally built in the West and Southwest, where there was an abundance of land, these sprawling beauties are the most expensive homes to build per square foot, because of the larger foundation necessary. Today, the term *ranch* has become a generic term for a one-story home. A raised ranch is basically a ranch with an elevated basement below, allowing for additional living in the lower level. The *new-ranch* popular today has elegant covered entryways with crisp columns and arched windows. They are one of the most popular styles being built because of their one-floor living advantage.

❀ *Southern colonial.* This plantation-style home evokes memories of Scarlett's beloved Tara. This two- to three-story home has stately columns along its large front porch. The porch was designed for outdoor living in the summer, providing shade and a perfect place to sit with a cool drink.

❀ *Split-level.* This style of home is defined by its split staircase that accesses two living levels about a half floor above each other. Sometimes it includes three levels (a trilevel home). It is economical and provides a large amount of living space by using the basement (foundation) as part of the finished living area. It is the only home style that can list the basement living space as part of its square footage for real estate value. (In real estate terms, any space below grade— (ground)—cannot be included in the square-footage count for value.)

❀ *Tudor.* One of my favorites from the exterior (usually too dark inside for my personal taste), it is modeled after an English country cottage. The contrast between its dark timbers and light-colored stucco is the focal feature of the upper half of this home. Steep pitched roof and brick for the lower portion finish this classic style. It is suitable for large- or small-scale homes, making it an ever-popular style. Inside its rich dark wood floors are matched with dark wood trims throughout.

❀ *Victorian.* Sometimes called Queen Anne, this style originated in Great Britain. Noted for its lavish details (gingerbread) and massive ornamentation, it was most popular from 1840 to 1900. It usually has a two-story wood frame with large rooms, high ceilings, and wonderful wraparound porches. With its high peaked roofs with ornately carved wood trim decoration, Victorian splendor is enjoying a renaissance. *Note:* If you would like to see some of the most incredible examples of Victorian style, visit Cape May, New Jersey—a particular treat at Christmastime.

The point to defining home styles is to help you in making good choices for renovating the exterior of your home. You have a choice when renovating to either stay with the existing overall style of your home or to completely change the style. All your selections should be based on which choice you have decided to make. For example, if you are planning on replacing your windows, it would make sense to complement the existing style of your home by choosing windows that are appropriate in style. To make windows most compatible with your home, pay attention to the type of window, number and pattern of window lights (individual panes of glass), and the size of the mullions and muntins (glazing or sash bars, which are the small moldings that separate individual panes of glass). It's easy to get so concerned with price, efficiency, maintenance issues, and delivery times that we end up sacrificing one of the most important issues—how they will affect the look of your home. Ultimately, any improvement you make should add value to your home, both aesthetically and financially.

Windows

Speaking of windows, let me elaborate on the subject. Changing or adding windows to a home is one of the most popular remodeling projects today. The national average cost for replacing windows is $5,976, based on replacing ten 3-by-5-foot windows with aluminum-clad windows. The resale value is estimated at $4,042. That's a 68 percent return. The primary reason for changing windows is for better functionality and energy efficiency. In other words, we want windows that operate easily and are also good at climate control. However, window remodeling projects can also enhance the visual beauty and livability of a room. Windows can be a very important part of expressing your creativity too. Replacing an old small window with a beautiful bay window can have dramatic effects.

Too often, I have been confronted with enormous windows that now require yards and yards of fabric to cover. This can add a lot of additional cost to your budget.

All of us seem to want bigger, airier rooms, and the easiest way to accomplish this is with windows. Architecturally, windows are a fundamental element of style both on the interior and exterior of your home. Choosing the right style of window is a big decision that should be considered not only for cost and durability but also for its permanent effect on the overall character and its relevance to the rest of your home.

It's important to also consider placement of your furniture when planning new windows. Too often, we forget to plan this part of the project when looking at windows. As with any other aspect of remodeling, it is key to a successful project. Here are some things to consider:

❀ *Be sure window placement and furniture arrangements work together.* This includes the elevation perspective as well. Elevation refers to a picture of your wall with windows and furniture drawn in. For example, a large piece of furniture can be balanced by adding a window above, or on either side of the furniture. The same is true for other architectural elements, such as a fireplace. Actually drawing a picture of what the finished elevation will look like can be very helpful in choosing the right window style and configuration.

ELEVATION SHOWING WINDOWS COMBINED WITH FURNITURE

❁ Don't forget to consider window treatments (for example, draperies). I know you're dreaming of all that wonderful light that you will be gaining, but consider your privacy needs as well. Do you really want the neighbors watching you sitting in your pj's while you drink your morning coffee? Also remember: That wonderful source of sunshine in the daytime is a big black hole after the sun goes down. Do you really want a 60-by-72-foot black hole as part of your interior decoration at night? If you are planning a window treatment with valance, be sure you have left enough wall space above it to accommodate this installation. Also be sure you have enough room on either side of the window for opening the draperies and still revealing the window.

❁ When positioning windows leave adequate space for the height of your furniture. Be sure the window sill height works with the height of the table you were planning on placing in front of it.

❁ Consider sun exposure and how it will affect furniture, fabric, and floors. It only takes twenty-four hours of sun exposure to remove color. I have seen a dining table that was set with place mats ruined by sun exposure. The position of the place mats is forever obvious because the sun faded the wood around them, leaving a distinct shadow of the place mats. Carpet and fabric, even leather, can change color and/or rot as a result of sun exposure.

Sunlight entering a room actually has a different color, or character, depending on what kind of exposure the window has. For example, A southern or eastern exposure will provide a warm, yellow kind of sunlight that is cheerful and comforting. Whereas a northern exposure will provide a more indirect sunlight that is cooler and whiter in color. Sunlight entering from a western exposed window will be harsher, with a sort of gray-white undertone. Recognizing these differences in advance can be helpful when you are planning color schemes for individual rooms. Obviously colors will change in appearance, depending on the kind of light they are exposed to. If you desire a warm and cozy room but the sunlight exposure is northern, or western, you may want to consider using warmer color tones in your decorating scheme. Yellow, red, and coral tones can add a lot of warmth. If on the other hand you anticipate a room will be too hot because of a southern exposure, you can cool it off by using cooler color tones such as blue and cool green shades for decorating.

A well-insulated window can make as much as a 20 percent difference in your heating and air-conditioning bills. A window should be tightly sealed and energy efficient. One study found that as much as 60 to 70 percent of total heat loss can be attributed to windows. As a result the National Fenestration Rating Council now offers a standardized method of comparing heat transmittance (*U-factor*), and provides consistent labeling systems for all glazed products, including doors. Each state has determined the maximum allowable U-factor for its particular area. Check with your state board for your specific U-factor numbers.

R-value is yet another energy-efficiency rating system. This rates the effectiveness of the insulating property of a material. The higher the R-value, the better it insulates. The average single-paned window has a resistance to heat flow of .9 R-value. A double-paned window has an R-value of approximately 1.7.

Solar-heat gain coefficient is a measurement of heat gain through glass from the sun. The lower the coefficient number, the lower the solar-heat gain. This is a particularly important number for colder cli-

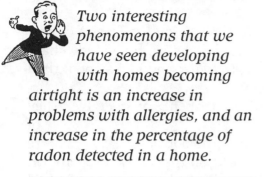 *Two interesting phenomenons that we have seen developing with homes becoming airtight is an increase in problems with allergies, and an increase in the percentage of radon detected in a home.*

mates, where there can be a benefit in gaining heat from passive solar elements.

UV blockage is a term that refers to how well a window reduces the ultraviolet transmission. This is most important when, for example, you are using a skylight. If a skylight is positioned such that it allows direct sunlight through, it will be a problem for furniture and flooring exposed to the sunlight. A laminated glass with a good UV-blocking rating will help deter the transmission of ultraviolet rays.

A factor affecting the insulation property or value of a vinyl window is whether or not it is insulated. By using urethane foam as an insulation, you can increase the insulating property immensely.

Most new windows do no need storm windows because most new windows are *double-glazed*. Double-glazed windows have two panes of glass with a vacuum between them for insulation. If you are really weather conscious, you can get triple-paned.

Another option would be *Low-E glass*. Low-E glass has a special coating applied that cuts down on the amount of heat transfer and also the rays that fade furniture and flooring. This coating reflects heat to the outside in summer and inside during winter. Pella® windows also have a special Low-E window that also includes a gold-tone Slimshade blind that reflects heat back into a room in winter. It can also be used to keep out unwanted sun and heat in the summer.

When our homes had leaky windows and doors, both the radon gas and the elements that caused indoor allergies had a means of escaping. Because now they are not able to leak out, we are having more trouble with them. If either of these issues is a problem for you, then consider getting professional advice on a ventilation system designed to address such issues. There are companies that specialize in radon remediation and also in designing filtration and ventilation systems for allergy sufferers.

WINDOW BASICS

❀ Windows basically come in two styles: sash (no frame) or pre-hung (with frame). A prehung also includes the jamb and side channels. If the frame around your window is sound and looks good, then consider using a sash replacement only.

❀ Windows are either fixed (do not open), or operable (they open).

❀ *Awning*-style windows open out at the bottom.

❀ *Casement*-style windows are hinged on the side and open outward like a door. With casement-style windows, the screen in mounted

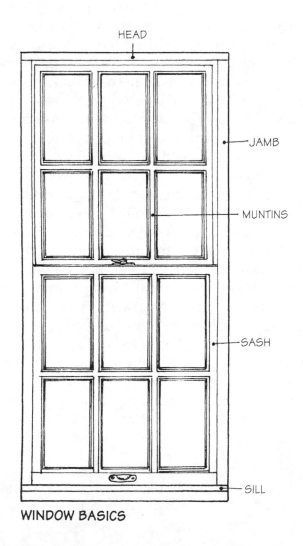

WINDOW BASICS

on the interior. Pella® recently introduced a casement-style window with a roll-down screen. This is, in my opinion, one of the neatest ideas to come along in a long, long time—especially since with casement-style windows, the screens are positioned on the interior of the window. It is really nice to be able to roll the screen up like a shade and get it completely out of view. It's called Rolscreen. I have casement windows in my home; I chose them because they were the best "look" for the overall architecture. However, I must admit, they are not my favorite style of windows because I like to be able to open windows when the weather permits. Casement-style windows are not the best style choice for letting in fresh air. If the breeze is blowing against the open casement panel, it can take a beating (making a lot of noise as well), and the fresh air cannot get in with the wind blowing against the panel. If on the other hand the wind is blowing with the panel, you will get the air in, however, it will be only in one direction. Consider how often you realistically open your windows. Most homes today are air-conditioned, and as a result, most windows seldom get opened.

❀ *Double-hung* windows open at the top and the bottom, sliding vertically on the sash. The screens *are mounted on the exterior.*

❀ *Hopper* styles are designed to tilt inward at the top. These are most often used in basements.

❀ Glass types can vary from window to window also. Older homes have what is called *sheet glass.* Two newer glass types are *tempered* and *safety.* When broken, tempered glass breaks into tiny beads. Safety glass has a thin layer of film in between two layers of glass. This keeps the broken fragments from flying. Low-E glass (see above) has a thin coating made of metal oxide that helps block certain types of solar rays. There are also several types of *frosted* or *obscure* glass designed to allow light in while providing a modicum of privacy. *Insulated* glass is two sheets of glass with a vacuum-sealed air pocket between for insulation. *Gas-insulated* glass has either krypton or argon gas between two panes for even better insulation. Argon gas is a clear, inert, natural gas that is denser than air. *Float* glass is the most common of all glazing materials for single-pane or insulated windows because it has a minimal optical distortion and is durable, scratch resistant, and relatively inexpensive.

All windows are available in wood, vinyl, or vinyl clad. The advantage to vinyl or vinyl clad is the fact that they don't need painting or caulking. Does this make them a better window? That is a matter of opinion. I believe that you will find a larger selection of window styles and sizes with greater architectural detail in wood, rather than vinyl or vinyl clad. Overall, vinyl-clad windows are less expensive than wood windows. One reason is that no on-site finishing is necessary. If you are choosing vinyl windows look for the seal of approval #101/I.S.2-97 from the AAMA (American Architectural Manufacturers Association). This will assure that the windows meet a uniform standard for performance and energy efficiency. Actually this seal represents performance testing for all windows, whether they are vinyl, wood, or aluminum.

Wood sashes and frames, some will argue, are better at insulating as well. Of course, there are different qualities to both vinyl- and wood-constructed windows. A good quality wood window should be made of kiln-dried wood (wood that has had the moisture removed by heating it in an oven). Pella® brand uses kiln-dried western pine. In addition, Pella® treats each piece with a water-repellent preservative that protects against decay and insects. They also have a special tough coating paint system that resists chalking, fading, and corrosion. All of these issues should be considered when choosing a wood window.

When checking the quality of the construction of a window be sure to check the way the sash is joined together. Sash corners should be carefully fastened. A quality window will have glued joining, as well as metal fasteners, and interlocking joints for maximum strength. Weather stripping is important for eliminating drafts and providing energy efficiency. A continuous strip that is welded at the corners is preferable.

CUSTOM VERSUS STOCK

There is so much variety available from standard or stock-size windows that it is generally not necessary to use custom-made windows. Most window dealers have dozens of shapes and sizes readily available. You can find sunburst, half-rounds, squares, quarter-rounds, and many others. By combining different shapes and sizes you can create a one-of-a-kind look without the cost of custom-made.

If you cannot find what you need in a stock window, then custom is always an option. I recommend using a well-known manufacturer.

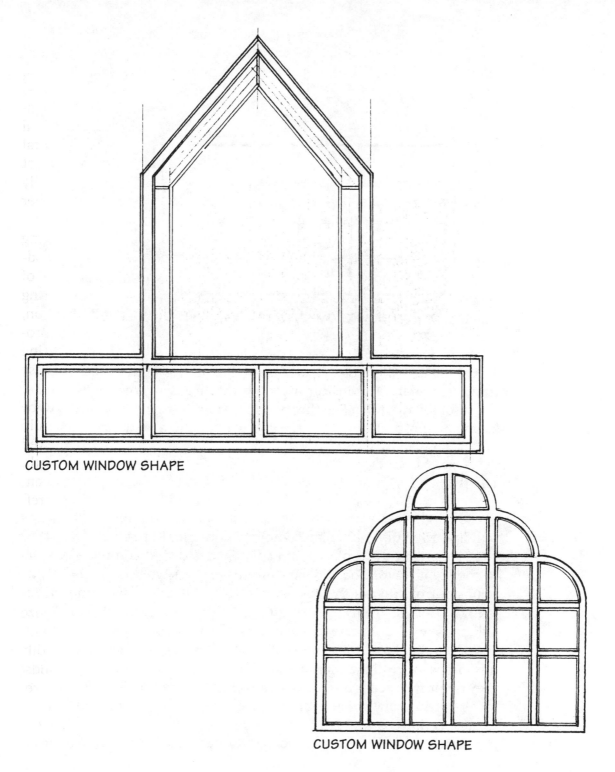

CUSTOM WINDOW SHAPE

CUSTOM WINDOW SHAPE

STOCK WINDOWS COMBINED TO MAKE
CUSTOM-LOOK

They will have the most experience and expertise in this area. The last thing you want is a custom window that leaks, warps, or just plain doesn't look right. Having the reputation and integrity of a larger, well-known manufacturer will assure you quality and a warranty.

Skylights

Skylights are wonderful for bringing the outdoors in. When I was building my home, I walked onto the construction site the day they were closing in the roof and immediately decided to add two skylights to the great room. The sun exposure from the skylights is on the side opposite from that of the windows. As a result, I now have the advantage of sunlight from two directions. As the sun moves around, I am able to take advantage of it, regardless of position.

Skylights have some other distinct advantages: They can provide five times more light than a sidewall window of the same size, and they don't sacrifice privacy. They can admit sunlight in winter from the lower-setting sun, providing much needed solar heat gain, which helps to cut heating costs.

Like windows, you can choose from operable- or stationary-style skylights. An operable skylight will allow you to bring fresh air in. If you decide to use an operable skylight, be sure a screen is available. Otherwise, you will also be inviting in some unfriendly insect friends.

Many times, the ceiling surface will be several feet lower than your roofline. This will require building a shaft to the skylight. You have a couple of options here:

❀ Straight-shaft skylights use the same size opening at both the roof and the ceiling. They are connected by short walls. Effectively, they work like a spotlight, guiding the sunlight to where you want it.

❀ Flared-shaft skylights use a larger opening at the ceiling than at the roofline. The advantage is a larger light beam spread, which provides light to a wider area.

❀ Vaulted-ceiling skylights have a much shorter shaft because the distance from ceiling line to roofline is much shorter. This allows for a lot more light entering the room.

❀ Solatube is yet another style of skylight designed for places where standard skylights can't go. They can be installed on any roof type, and are about one-third the cost of a standard skylight. They use a rooftop reflector that collects and directs light through a mirrored cylinder to a diffuser in the ceiling. A 10-inch-diameter model admits more light than a 2-foot-square box skylight. For more information, call Solatube at 800-773-7652.

One of my favorite decorating tricks for skylight shafts is to mirror the surface of the shaft. The mirrors reflect the sky, giving the impression of a much larger opening. It's a great look!

Just as with windows, take into consideration sun-shading options. If you live in a warm climate, consider window tinting to block harmful sun rays. You may also consider built-in shades or blinds. If you live in a colder climate, be sure you have good weather stripping.

Shed-Roof Dormer

This basically is the use of windows for a portion of your roof. They are generally constructed as shed dormers. A shed dormer is generally larger than a peaked gable-type dormer. However, here are a few words of caution from my own experience. If you are installing them on a southern-exposed side of your home, it will get HOT in this room—very HOT! In addition, the warranty on the seals (double- or triple-glazed windows are sealed with air gaps between) is *not* warranted if installed at an angle. If you are using them as a roofline window, this is an angle, and they are not warranted. Why? Because obviously the seals will leak in this particular type of installation.

When this happens your windows will get cloudy, because they will allow moisture in, which will fog them up. Eventually, all double- or triple-glazed windows will have a seal leak and develop this problem. However, under normal use (as a standard sidewall window) they have a standard warranty for a specific time period. Unfortunately, as a roofline, shed-dormer-style installation, they do not have a standard warranty on the seals. This can be not only an unsightly problem but also an expensive one to fix. Your only choice is to replace the windows.

Another word of caution: I recommend aluminum cladding on the exterior surface of roofline windows. They are subject to some of the most extreme weather conditions, both cold and hot in nature. If the frame structure surrounding the windows is not maintained properly, you will have leakage problems. I suggest semiannual inspections on the windows, framing, and caulking for roofline windows.

Exterior Doors—Don't Slam That Door!

The sound of a screen door slamming is almost nostalgic. Growing up in a family with six children, the sound of a door slamming was nearly constant, as was my mother's "Don't slam that door!" It makes me smile to think of it now. Perhaps it's because today's doors generally don't slam. Instead, they have sophisticated door closers that control the speed of closure and prevent little ones' fingers from getting pinched.

Nonetheless, doors are one of the hardest-working elements in your home. They provide security and protect us from the harsh elements of Mother Nature.

They also create the first impression most people have about your home. The character and condition of your door are a direct reflection on you and your family. When was the last time you walked up to your front door and really looked at it? If it's been awhile, try it. After a winter storm during the holidays last year, my front door suffered some bruises from a holiday wreath. Because my normal method of entrance is through my garage, it took me a few months to notice. Repainting it was one of my first spring projects this year. Speaking of paint, a front door is one place you can really express yourself. It is the perfect opportunity to get a little wild with color and accents such as wreaths and planters. For one of my clients, we painted the front door a wonderful shade of periwinkle blue-purple. It was the perfect introduction to the interior of her contemporary, brightly colored interior.

Most exterior doors are made of solid wood or an insulated foam core covered with steel or fiberglass. I prefer the look, feel, and sound of a solid wood door. The traditional essence appeals to me. However, foam-core doors do have their place, particularly in colder climates. They can be better insulators, and they definitely require less maintenance. Steel doors (which I now have) are much stronger and provide better security. Fiberglass-clad doors are more expensive, but lighter in weight and are the most maintenance-free doors available. Some fiberglass doors can actually be stained to look like wood.

When choosing an exterior door, be sure to also take into consideration security. A recent trend is the use of more and more glass around entry doors (as sidelights), as well as glass-inserted doors such as leaded-glass panels. This is a great look, but it can pose a security problem. Most building codes will require that break-resistant tempered glass be used for all glass doors, including French patio style, or sliding doors.

Over the last five years, the trend has been to use French-style patio doors instead of sliding doors. I must admit, I agree with this preference. French-style patio doors just look and feel better to me. Sometimes there just isn't the interior room available to accommodate the patio style. Sliding doors definitely have the advantage here. In either case, get the best you can afford. Doors are expected to take a lot

FRONT DOOR TREATMENT

FRONT DOOR TREATMENT

of use and abuse. The better the quality, the longer they will perform without trouble and the better they will look over the long haul. To help keep your exterior doors looking nicer longer, protect them with an overhang or porch to shield them from the elements.

Screen/storm doors can also provide protection and additional ventilation in nice weather. In choosing a screen door, be sure its character matches the character of the rest of the home. Also be sure you like the way it looks with your entry door. After all, what is the point of choosing a beautiful entry door and then covering it up with a cheap screen door? Wood-framed glass-insert storm doors are both beautiful and effective. The larger expanse of glass (or screen) will allow the beauty of your front door to clearly shine through.

Exterior Finish Materials—Paint, Stain, Stucco, or What?

PAINT AND STAIN

Choosing the right finish material for your home has many facets. Obviously it should improve the appearance of your home, but it should also make your life easier with less maintenance. Two of the most popular products are *paint* and *stain*. However, most people don't really know the difference between the two. Stain is basically a watered-down version of paint. Which means that it offers very little protection. As a result, if you are intent on the wonderful rustic look of stain, then you need to be prepared to redo it every couple of years. Paint on the other hand will provide a water-resistant finish that can last eight to ten years.

Exterior stains are available in either semitransparent or opaque finishes. Semitransparent stain has a flat finish that allows the grain to show through. Opaque stains have a texture similar to paint and will cover most of the grain, while still allowing the texture of the wood to show. Opaque stains also provide better, longer protection than semitransparent stains. Stains look best on rough-textured wood such as redwood or cedar. My home in New Jersey had a cedar shingle exterior that I stained with semitransparent stain. It was the perfect look for a "shore" home—rustic and natural. A word of caution: When using a semitransparent stain, be sure to test it for color first. Some woods, such as cedar and redwood, can actually change the color of the stain. By testing first, you can adjust the color accordingly. I chose semitransparent because opaque stains can peel. This is because they are thicker but not thick enough to block UV light. As a result, within a very short period of time, the UV light can degrade the underlying wood fibers and cause the opaque stain to lift off. The American Plywood Association found in a recent study that using a "primer" is critical not only to a long-lasting paint finish but also to an opaque stain finish.

The most important thing to know about paint brands is that they are chemically specific. What this means is that paints and their counterparts, primer, are chemically designed to work together. You should not mix one brand of primer with another brand of paint. It is the combination of the primer and the paint that provides your home protection against water and ultraviolet light. The binder in the paint seals out the water, and the pigments block the UV light.

Paint is a wonderful product that can be used easily over wood or cement. Paints allow for a variety of finishes such as flat, satin, or glossy. The glossier the finish the easier it is to clean.

There are basically two types of paint—oil based and water based (latex). Oil-based paints create a moisture-block layer. As a result, if the surface you are painting is not completely dry, then you will be setting yourself up for a disaster. The wet surface below the paint will prevent the paint from sticking. As a result, you will get peeling, cracking paint. Latex paint on the other hand is not impermeable to water, so it is not as critical that the surface be dry.

SIDING

At one time, aluminum siding was "state-of-the-art." Not anymore—today anyone considering an alternative to wood siding should choose vinyl. Why? Because unlike aluminum, the color in vinyl siding permeates all the way through. It is also more flexible and resilient to denting. If you are considering siding, it is also a good time to consider adding insulation. You can choose having insulation blown into the wall or installed in rigid panels behind the siding. In most cases you can install the new siding right over the old, provided the old is in

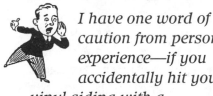

I have one word of caution from personal experience—if you accidentally hit your vinyl siding with a Weedwacker cable it will cut through the siding.

relatively good condition. This can save you a lot of money. A layer of rigid insulation should be installed between the old and the new.

Adding vinyl or aluminum siding can bring a great amount of return on investment according to *Remodeling* magazine's Cost & Value report. The national average cost of a 1,250-square-foot siding project was $5,099, with a resale value of $3,593. That's 71 percent of return of cost.

HardiPanel® is a relatively new product on the market. It is a fiber-cement siding that is an attractive alternative to wood siding. It has the true look of wood with the durability of cement. It is resistant to damage from humidity, rain, snow, salt air, and termites. It is dimensionally stable and will not crack, rot, or delaminate. The manufacturer is so convinced of the durability of this product that it carries a fifty-year transferable product warranty! I think that's pretty amazing. It comes primed and ready for painting. The recommended finish is 100 percent acrylic latex paint. It is available in four different wood textures: smooth, cedarmill, colonial smooth, and colonial roughsawn. Hardi*Panel*® is available in both vertical and horizontal panels. This same company makes Hardi*Plank*® and Hardi*Soffits*®. Installation is similar to wood, including the cutting, fastening, and finishing. It can be cut with a circular saw and a carbide-tipped blade. I think these products are some of the best new ideas in many, many years. It has the richness of wood with incredible durability and low maintenance. The manufacturer is James Hardie.™ For more information, call 800-9-HARDIE.

My style preference for siding is seamless products with a graining that gives the rich look of wood.

STUCCO AND DRYVIT

I love the look of stucco, but I hate the maintenance because it usually ends up dirty and cracked, and it's impossible to repair without it looking obvious. The good news is there is a wonderful alternative that not only provides you with the "look" of stucco but also insulates your home to keep the temperature more consistent year-round. If I sound like a commercial, I'm sorry, it's just that I really like Dryvit.® The Dryvit® system combines insulation boards with a composite material that's thermally efficient. The system's insulation boards are installed over a building-code approved weather barrier. It can be installed over plywood, OSB substrate, or existing siding. Then a textured finish is applied. This special finish is available in a wide range of textures and colors. They also have a product called ULTRA-TEX®, which can create the look of stone, tile, slate, and brick for both horizontal and vertical applications. This system provides as much as an R4 insulation value. I first used this product on my passive solar home in 1981. At that time there were only a few colors and texture choices available, so I had them custom-color a finish for me. Since that time they have developed a wonderful variety of not only textures and colors, but also architectural details, such as window borders, arches, railings, columns, and more. The finish also has a dirt-resistant chemistry. This helps to keep the colors bright and clean. On my current home, I have a combination of vinyl siding, brick, and Dryvit.® The combination of textures creates an interesting and eye-appealing look.

BRICK

Brick is an American tradition. Brick evokes the feeling of old-world craftsmanship. Its aesthetic appeal, and flexibility in design, give it great advantage for building. Brick comes in a variety of shapes, styles, finishes, and sizes. My favorite is handmade brick. Each brick is individually formed in a wooden mold creating a wonderful texture. Other choices include: *extruded*, or stiff mud, which are wire cut; *molded*, or machine molded with a sand finish; *glazed*, with a shiny seal that is applied before firing, making it an integral part of the brick; and *clay coat*, a smooth nonglaze finish.

Brick is available in standard $3^5/8$ by $2^1/4$ by $7^5/8$ inches or an oversize of 4 by $2^3/4$ by $8^1/2$ inches. With the oversize brick, you use less brick, which also reduces the installation time because you need less mortar, which usually means less cost. You can also choose "thin brick." It is only $1/2$ inch thick. This reduces the weight as well as the cost.

BRICK HOME

Brick can last hundreds of years. Here in Lancaster County, Pennsylvania, there are many century-old homes with their original colonial brick. The mortar on the other hand will not last hundreds of years. Crumbling mortar is a problem that needs to be addressed. If not, the brick will come tumbling down. The mortar is what holds the brick in place. *Repointing* is the term used for the process of chiseling out the old mortar and replacing it with new. After chiseling the old out you should clean it with a spray of water to remove any debris. Repointing is very popular here in Lancaster. It is amazing how new a building can look after this process.

When choosing your mortar color, recognize that it will affect not only the overall appearance of the color of your brick but also will be a determining factor in the amount of definition and/or unity of the overall finish. A contrasting mortar will create a more textured (checkerboard) effect. A complementary or matching color will create a smoother appearance. I prefer taking clients directly to the brick showroom, where they usually have most of the brick choices mounted on boards with various color mortar options. Brick is a big investment; you don't want to spoil it with the wrong color mortar. Also, brick is the most costly finish material available because of the labor and time involved to install them.

CULTURED STONE

Cultured or man-made stone is a popular choice for remodeling. It is less expensive than real stone and easier to install. Cultured stone is cast in molds that were made by using real stone as a model. This allows even the tiniest details to come through in the mold. Using a blended cement with aggregate and iron oxide pigments, it looks and feels like the original stone with half the thickness. As a result, it is light enough not to require additional foundational support. Most cultured stone carries a thirty-year warranty that includes colorfastness.

It is my opinion that the only color choice for stone mortar is a color that blends into the stone. This way the stone is the focal feature and not the mortar.

It can be installed over a wood-frame surface or masonry or concrete. It is adhered with mortar and permanently installed.

Most manufacturers of cultured stone also make complementary architectural trim elements, such as window and door trim stones. Cultured brick can be used indoors as well as outdoors. I often use it for facing an interior fireplace. It can also be used to build landscaping or garden walls. It is available in a nearly endless choice of stone styles, shapes, and sizes. Just as with brick, the color or mortar you choose is critical to the overall look.

Garage Doors

Here is just a short note on garage doors. They are one of the largest elements on the exterior of your home, covering as much as 40 percent of the exterior. Particular attention should be paid to them so that they complement and not detract from the overall appearance of your home. If you have a home that would look better with a historic-looking style, rather than the more contemporary upward-opening doors, don't fret. There are some wonderful options of garage doors that have the appearance of the old-fashioned carriage or barn-style doors that seem to open outward. However, in fact, they open *upward*, with a standard remote control. Clopay's Carriage House series (800-225-6729), and Hahn's Woodworking's farmhouse tongue-and-groove arched top doors (908-241-8825) are two good options to check out.

There is also a new garage-door opener that uses a *rolling-code* technology. Each time you use it, it changes the code. The advantage to this new system is that it creates an unlimited number of code configurations. Because the old system had a limited number of codes nationally, it was easy for thieves to detect which code was being used. Most new garage-door openers use this new rolling-code system.

Some of the more expensive garage doors now have a lifetime warranty, which means overall they can end up costing less than a less expensive brand. What makes a garage door better is its insulation value for both sound and temperature. This is important not only for keeping the noise level down but also is even more important if you use your garage as a workshop or exercise area. It can also affect the temperature of the inside of your home. If you have one or more common adjoining walls from the interior to your garage, an inexpensive, less-insulated door can dramatically affect the home's interior temperature. A door with a thermal break between the inner and outer steel skins is the best insulating choice. The insulation value can range from R-6 to R-17. Of course, the more windows on the garage door, the less insulation or "R value" you will get. Most garage doors use polyurethane foam as an insulator. This can either be blown into the interior of the door, or glued onto the back.

Other important features of quality to look for are: doors that are 24- to 26-gauge steel thick. The lower the gauge number, the thicker or getter the steel. Other good options for door material choices are polyethylene or HDPE composite board. HDPE is extremely durable and will withstand most impacts well. (This would have been a good choice when my brother was a teenager and forgot to "open" the garage door before attempting to drive out!) HDPE is a textured composite material that can be stained to match your home.

A new safety design in garage doors is "pinch resistant panels." These are designed to protect little fingers by virtually pushing little fingers out of the joint as the door closes.

Words of Wisdom

❀ Another pet consideration is their uncanny ability to damage screens. Some screen manufacturers now claim to have a *stretchy* screen that is more resilient to such trouble.

❀ When choosing windows and doors, consider your family pets. They love to sit and look out to see the world. By placing windows lower, they can easily enjoy this treat.

❀ A drafty door can be easily fixed by installing a door sweep weather stripping. It's a rubber strip that attaches to the bottom edge of the door. I will warn you, though, many a pet (both cat and dog) find this a nearly irresistible chew toy!

❀ If you live near the ocean, be sure your door and window hardware are designed especially to resist the corrosive effects of salt air and mist. Pella® makes a line of hardware called *Seacoast* specifically for use in such conditions.

❀ Don't stop at the window, go outside. Would adding shutters or awnings improve the overall appearance of your home? If so, be sure to take into consideration the placement of shutters when you are installing your windows and leave room on either side to accommodate them. Also, think about adding window boxes to the front of your home. This simple idea can add a lot of interest and beauty.

❀ If you have high ceilings, consider installing an eight-foot-tall entry door. It will work much better with the proportions of the rest of the home.

❀ The most popular window choice in the East is aluminum-clad units. Demand for new windows is weakest among home buyers in the Midwest.

❀ A paint or stain will stick better to a newer surface than a weathered one. If you want your finish to last, give it a good sanding first.

❀ The most popular colors for the exterior of homes are still neutrals, with white being the number one choice.

❀ If you are planning on installing your own new garage doors, then select a door with a safe drill-adjustable counterbalance spring. Otherwise, you will find yourself with a broken spring.

11

OUTDOOR LIVING

Over the past ten years, more money has been spent on living outside than ever before. Why? Because we like being outdoors. In fact a recent poll found that the most used space of our home is now the "outdoors"—whether it's a porch, patio, sunroom, or garden.

Sunrooms and Conservatories

Traditional Home magazine found in a survey in 1997 that sunrooms are one of the top three home amenities desired by women and top six by men. And not only are they aesthetically pleasing by creating an environment of sunshine, but they also make financial sense. According to the National Association of the Remodeling Industry, they can yield anywhere from a 70 to 110 percent return on investment.

One of the biggest surprises to most homeowners is how much time they actually spend in their new sunroom. In fact, most people spend more time there than anywhere else in the home. What started out as a place to find a little peace and relaxation ends up becoming an important component of the home. This in part might be because we *feel* better in a sunroom. This isn't just a figment of our imagination. It is a scientific fact that our bodies need sunshine. What better place to get it than in the comfort of your own air-conditioned space?

Do plan to make your sunroom a three- or four-season room by properly ventilating, screening, and heating it. If you don't plan properly, you can end up with an *oven* instead of a sunroom. Using ceramic

SUNROOM ADDITION

SCREENED-IN PORCH ADDITION

tile, tinted cement, or any hard-surface material for the floor will not only last longer under severe sun exposure but can also add a passive-solar heat gain for those cold winter days.

A trend gaining popularity again is the use of *conservatories*. A conservatory traditionally speaking is a greenhouse attached to your home. During the opulent period before the Depression, they were considered essential to the homes of the wealthy. Well, they are not just for growing plants anymore, nor are they reserved for the upper-crust rich. A conservatory is a beautiful way to add the splendor of nature into a sun-drenched garden room. Conservatories offer a distinct style that cannot be accomplished any other way. They are custom-designed specifically for each home. In addition, each project is custom tailored to meet the specific climate conditions for your area. Renaissance Conservatories here in Lancaster, Pennsylvania, have developed and incorporated some pretty incredible details that make their conservatories not only more durable but also much more functional. They have a positive weep system for sloped and vertical glass, concealed structural steel supports, a "hideaway" venting ridge system, motorized roof windows with rain sensors, double-sealed A-rated insulating glass, and maintenance-free roof capping materials such as aluminum, copper, lead-coated copper, brass, and zinc. I learned from personal experience just how important these features can be. I ended up with an awful "weeping" problem that eventually rotted out the wood structure. Although mine was built of redwood, the preferred wood choice, it was not kiln dried (oven dried). This too contributed to the problem.

If you are thinking about a sunroom, give serious consideration to a conservatory. They are perfect places for entertaining, living, as well as a great place for a hot tub. Architecturally they are incredibly beautiful and can be designed with the traditional classic look of a greenhouse or given a streamlined contemporary style. I had a contemporary style on my last home.

Whatever your choice, be careful in choosing all finish materials and also your furniture. Be sure they can hold up to the harshness of the sun, which can within a mere twenty-four hours completely remove color from fabric, wood, and wallpaper. There are fabrics specifically designed to withstand such sun exposure. Check with the supplier before purchasing. Be sure you have a written specification sheet saying that the products you have chosen meet the criteria for long-term durability in a sunroom.

INSIDE VIEW OF CONSERVATORY

OUTSIDE VIEW OF CONSERVATORY

ANOTHER STYLE CONSERVATORY

Porches and Decks

Last November I decided to build a screened-in porch. It was truly one of the best renovating choices I ever made. It has made this summer wonderful. Since I rarely take a day off, I treasure my mornings and evenings spent on the porch. I am delighted with the building materials I chose. They are providing a care-free and maintenance-free living space.

Since my yard slopes, the porch was designed as a deck-style on stilts rather than with a foundation. I wanted the look and feel of a wood deck floor, without the splinters, hassles, and warping problems. I discovered a relatively new product called Trex. Trex is actually a wood-polymer that is manufactured to look like lumber. It will weather gray with age, or it can be stained, just like real wood. It does not need to be sealed, and it is also warp free and won't rot or get damaged by insects. The rest of the porch materials are aluminum and vinyl. I used aluminum framing for the screening and vinyl railings. The screen sizes have been designed so that I can add glass windows later, to make it a three-season room.

Another new product in the outdoor wood area is a new pressure-treated lumber with no arsenic or chromium preservatives. Unlike most pressure-treated lumber, this one is safe for the environment. It uses an alkaline copper quat. Quats are fungicides that attach decay organisms. It is the same product used in swimming pool chemicals, shampoos, and hospital cleaning products. Instead of turning gray with weather, this new product will turn a warm brown color. This product is manufactured by Chemical Specialties, Inc. For more information, call 704-522-0825.

Decks are more popular than ever. They have also become much more stylish, with multilevels, gazebos, and lots of wonderful gingerbread trim and details such as a lattice railings. The national average

MY SCREENED-IN PORCH

construction cost for a 16-by-20-foot deck is $5,927, according to *Remodeling* magazine's Cost & Value report. They also found that homes with a deck or porch sell faster. Here are some design tips that have not only a visual impact but a practical one as well. If building codes permit, use a built-in bench instead of a railing. To integrate the deck with the rest of the house, paint the band boards, railings, posts, and balusters to match your existing siding.

One area often neglected in planning a deck is the area below and around it. It is particularly important to consider landscaping and screening when building an elevated deck. Painted lattice is an easy way to not only finish off the space below decking but it also has the added benefit of giving you a place to hide some outdoor equipment. Plan ahead to add an access door in your lattice panels. If your deck is on a slope or at a height that will allow, a popular feature in my neighborhood is to add a patio below. This can be a great place to hang a swing mounted to the underside of the deck. It can also be a great place to add a cascading waterfall leading to a pond. Use your imagination to incorporate the deck into a complete backyard garden.

Redwood garden shelters are another way to attractively detail a deck while adding some much-needed shade. By adding a trellis to your deck, patio, or garden, and planting your favorite vertically growing vine, you can provide a beautiful living shelter. There are plenty of do-it-yourself kits available. The California Redwood Association is a good place for ideas. Web site: www.calredwood.org).

Pergola is a word you may have heard recently. It's not a new word, just an old idea that has come back in style. Basically it is a fancy trellis—an arbor formed of horizontal trelliswork or a colonnade having the form of an arbor. Trellis patterns can be simple or complex, with overlapping square patterns and accents of circular cutouts, acting as windows. A pergola is usually attached to the house and is a great way to connect a detached garage to the house.

Gazebos are extremely popular. They evoke a sense of romance while providing the perfect respite from the hot sun. By enclosing a gazebo with screening, you can create your own personal outdoor room.

Retaining or garden walls are another important feature to the foundation of an outdoor living space. They not only hold back the earth but they also add interesting architectural details, like terracing and patio definition. Building a retaining wall requires some engineer-

BEFORE LANDSCAPING VIEW

AFTER LANDSCAPING VIEW

USING A RETAINING WALL IN LANDSCAPING

ing in order for it to withhold the stress of earth and water pressure. They are strictly regulated by local building codes. Be sure to obtain all necessary permits before undertaking such a project. Whether built of stone, concrete or brick, a retaining wall requires good drainage. In addition soils vary in their ability to absorb or drain water. They also react differently when wet or frozen. As a rule of thumb, soils that drain well and remain stable when wet will put less pressure on the retaining wall. Gravel and sandy gravel are good examples. Whereas clay or silt will absorb more water and therefore put more pressure on the wall. Check with your local building department for information and advice on local soil conditions.

Outdoor Lighting

This is better known as *lightscaping*, because lighting has become such an integral part of our home exteriors. No longer just a lamp to light our path, lighting has now become a way of complementing and showing off your garden and your home. Floodlights and spotlights when used properly can be ideal for accenting outdoor areas. Lighting literally can bring the outdoors in by expanding the interior space of your home at night. Consider how lighting will, for example, be seen from your dining room. Do you have a tree or architectural element that could be interesting if enhanced with lighting? Pay particular attention to trees, shrubs, and flowers.

Artificial light, unlike natural sunshine can be controlled. You will be amazed at how interesting a tree can look when lit from underneath instead of from the top (with the sun). Texture, size, even color can be completely changed with the right light. Consider light as a color when planning your lighting scheme. For example, fall foliage is better under a yellow/red hue. Also be careful if you are using ground lights. I have actually seen fall foliage (leaves) catch on fire. They had fallen into the ground fixture. It didn't take long for them to begin smoking!

You're Cooking!

The trend toward outdoor kitchens may have started in Florida and California, but it is becoming a national phenomenon. To attest to this, just about every kitchen stove company now has a line of fancy outdoor cooking appliances, including Jenn-Air, General Electric, Thermador, and even the familiar Weaver grill company. The basic outdoor grill isn't so basic anymore. They now have such amenities as side burners, shelves, and deluxe wheels. This is just the beginning. Of course, we all need the accompanying deluxe tool sets, rotisserie kits, broilers, woks, and wire-cleaning brush tools! The fastest growing category of gas grills is in the $400 to $750 range.

We love entertaining outdoors, and what better way to entertain than with food? It is this combination of the best of best that makes this so attractive. Some builders are even offering packages that include a fully equipped outdoor cooking area. A built-in gas grill with sink set into an island costs in the range of $1,000 to $5,000, depending on your choice of features. Other interesting popular options are wood-fired pizza ovens and ice-cooled salad servers.

Of course, if we're going to be spending so much time, money, and effort for our outdoor entertaining, we better not take a chance on a too-cool-breeze stirring and ruining our plans. Many of these outdoor kitchen/entertainment rooms now include heaters—both electric ceiling-mounted, and/or wood-burning patio burners. These are often made of clay pottery. I've opted for an old-fashioned Lancaster County Amish quilt instead. If it gets too cold for the quilt, I go indoors.

Pools

The swimming pool is one of those amenities that brings to mind either the tranquillity of the ultimate luxury, or it conjures up work, maintenance, and trouble. This has a lot to do with the part of the country you live in, and your own personal desire. Here in the East, a pool can be a negative when reselling your home. Because pools can be used only three to four months out of the year they are considered to be expensive and time-consuming to keep up. This said, in many other parts of the country they can add $10,000 to $30,000 to the value of your home. In warm climates a pool can help sell a home faster—as long as it is in good condition and updated.

Before you rush out and sign on the dotted line, be sure you understand all the costs involved, including the "hidden" costs. A standard 15-by-30-foot pool could cost $25,000. Your property taxes will also increase. And like any other remodeling project, it's worth more if it has the latest technology and bells and whistles. Such things as special lighting, with the new fiber optics, and heat-resistant finishes on the surrounding patio to keep your tootsies cool are a positive investment.

Cascading waterfalls, fountains, and a moat with a continuously moving current for floating your raft are just some of the additional goodies being incorporated into the backyard pool. Real estate agents estimate that lush landscaping and other pool items can recoup as much as 80 percent from the investment.

Here are some tips when considering a pool. It is my opinion that this should not be a do-it-yourself project. Two families in our neighborhood built in-ground pools this summer. One chose to do it himself. He started two months ahead of the crew building the other neighbor's pool. The neighbor who contracted the crew had a finished, landscaped, beautiful pool within six weeks. The do-it-yourself neighbor, three months into the project, still had a lot of dirt and mud. It is obvious even at this point that his project is going to look homemade and not professional. A pool is an enormous undertaking. If you decide to do it yourself, be sure you are equipped.

Regardless of how you decide to proceed, here are some guidelines:

❀ Check with your local building code authority. Be sure you will have no problem obtaining a permit to build a pool. Such things as water tables and water flow can affect whether or not it is possible to build an in-ground versus above-ground pool.

❀ Be sure you are familiar with the required fencing codes for your community. This can really affect not only the design of the pool but also the overall cost of the project. Large expanses of fencing can get to be pretty expensive. Find out what the required "height" for pool fencing is also. The do-it-yourself neighbor found out *after* installing his fence that it is too short to meet code.

❀ Make a budget of what you want to spend. Be sure to include all costs, including landscaping, lighting, building permits, and poolside furniture and structures.

❀ Consider an automated maintenance system.

❀ Get at least three written estimates from reputable pool builders. Be sure they make an on-site inspection and determine what kind of equipment will be needed or can be accommodated in your yard.

❀ Always check references. Be sure the builder has the proper license and insurance.

❀ Check to be sure the builder is following the standards set by the National Spa and Pool Institute/American National Standards Institute.

❀ Ask for and check out the builder's warranty for the pool and additional equipment.

❀ Check with your home insurer to see what kinds of requirements can save money on your policy.

❀ Consider building a "safe" pool. A safe pool is defined as a pool without a diving board. They are generally no deeper than five feet. A safe pool will generally cost less to insure.

❀ Plan your season. If you want to actually be able to "swim" in your pool next summer, be sure you are scheduled for late fall or really early in the spring. There is nothing worse than spending your summer "watching" a construction site instead of "enjoying" the pool of your dreams.

Words of Wisdom

❀ Clean your exterior wood once a year. Use a bleach-free detergent. For mildew, employ a small amount of household bleach diluted with water, then rewash with a bleach-free detergent. Be sure it is completely dry before applying any wood finish.

❀ Protect your exterior wood with stain or clear finish sealer. New wood needs protecting too. That old wives tale about waiting a year before treating new lumber is NOT true. By treating new lumber you will actually reduce the splitting and warping and help it weather more evenly.

❀ Keeping landscaping looking neat is a real chore. Bark mulch looks good, keeps weeds in control, and helps retain soil moisture.

Warning: It should not be used next to the house because it attracts termites. An alternative is lava rock. It is available in pebble and nugget form from most home and garden centers.

❀ To prevent iron furniture from leaving a rusty footprint on your concrete, use rubber caps under the furniture legs.

❀ To remove mildew and dirt from outdoor cushions, wash with a bleach-and-water solution. Then let air dry.

HAVE FUN!

After all is said and done, what really matters is what's important to you. Keep your priorities straight and in line with your own personal values. Any remodeling or renovating project must make sense for you and your family. Don't worry about keeping up with the Joneses. After all, they don't live in your home. Realize that no home-improvement project will *make* money. You will never recover your entire financial investment dollar for dollar. It is true, however, that sometimes you will lose a lot more money by *not* renovating before selling your home. For example, the perceived negative value of an out-of-date kitchen can be greater than the actual cost of replacing the kitchen with a new mid-range-quality kitchen. A home with worn, dirty carpet and out-of-date kitchen flooring can really take a financial beating when you go to sell it. New carpeting and vinyl flooring can bring it up to par with the resale value of the rest of the neighborhood.

When considering any renovation project, look and plan to the future. If your children will be out of the nest in the next few years, then plan accordingly. Remember that fashions are constantly changing. Don't get caught up in a too trendy idea that you may live to regret later. Before signing on any dotted line, be sure you are comfortable with all aspects of the project, contract, and people you will be dealing with. If you are uneasy or feel pressured—then wait. Better safe than sorry.

Your home is your biggest investment. It is meant to provide you with shelter and a restful haven. Ultimately, the atmosphere in your home is the result of not just structure and finish materials but of your attitude as well. If you are not emotionally prepared to deal with the chaos of renovation, then don't do it. A happy family is not dependent on having the perfect home. Besides, there is no such thing as a "perfect home." No matter how much renovating or decorating you do, there will always be something else that could be done.

CREDITS

I have been very fortunate to work with many professional specialists. They have not only made my work easier, but they have taught me a lot. Below is a list of those who have agreed to be a part of this book by allowing me to include their expertise and/or photos of some of their work. I give them the highest recommendation and thank them for their help.

Interior illustrations by Melissa Burkhart

Photography by WM Productions, 128 E. Clay Street, Lancaster, PA 17602 (pages 135, 236, 237, 290, and 292)

Kountry Kraft Kitchens, P.O. Box 570, Newmanstown, PA 17073 (pages 71, 96, 135, and 155)

Kreider & Diller Builders, 1936 Columbia Avenue, Lancaster, PA 17602 (pages 24 and 25)

River Valley Landscapes, Wrightsville, PA 17368 (page 292)

Plain & Fancy Kitchens, P.O. Box 519, Schaefferstown, PA 17088 (page 98)

Media Rooms, Inc. 34 Whitetail Drive, Chadds Ford, PA 19317 (page 247)

Renaissance Conservatories, 1888 Commerce Park East Road, Lancaster, PA 17601 (pages 288 and 289)

Wheatland Custom Homes, Inc., 447 Granite Run Drive, Lancaster, PA 17601, www.hometrends.com/wheatland. (page 28)

Dutch Quality Contractors, 72 Pitney Road, Lancaster, PA 17602 (pages 79, 82, 197, and 286)

Gable Designs, 682 Central Manor Road, Lancaster, PA 17603, www.gabledesigns.com (pages 236 and 237)